Empowering the Mentor

of the

Preservice Mathematics Teacher

Empowering the Mentor
of the
Preservice Mathematics Teacher

Edited by

Gwen Zimmermann
Adlai E. Stevenson High School
Lincolnshire, Illinois

Patricia Guinee
Peoria Public Schools, Peoria, Illinois

Linda M. Fulmore
Mathematics and Equity Education Consultant
Cave Creek, Arizona

Elizabeth Murray
Navajo Elementary School, Albuquerque Public Schools
Albuquerque, New Mexico

Copyright © 2009 by
THE NATIONAL COUNCIL OF TEACHERS OF MATHEMATICS, INC.
1906 Association Drive, Reston, VA 20191-1502
(703) 620-9840; (800) 235-7566; www.nctm.org
All rights reserved

Library of Congress Cataloging-in-Publication Data

Empowering the mentor of the preservice mathematics teacher / edited by Gwen Zimmermann ... [et al.].
   p. cm.
  Includes bibliographical references.
  ISBN 978-0-87353-628-8 (alk. paper)
 1. Mathematics teachers—Training of. 2. Mathematics teachers—In-service training. 3. Mentoring in science. 4. Mentoring in the professions. I. Zimmermann, Gwen.
  QA10.5.E47 2009
  507.1—dc22

The National Council of Teachers of Mathematics is a public voice of mathematics education, providing vision, leadership, and professional development to support teachers in ensuring equitable mathematics learning of the highest quality for all students.

For permission to photocopy or use material electronically from *Empowering the Mentor of the Preservice Mathematics Teacher*, please access www.copyright.com or contact the Copyright Clearance Center, Inc. (CCC), 222 Rosewood Drive, Danvers, MA 01923, 978-750-8400. CCC is a not-for-profit organization that provides licenses and registration for a variety of users. Permission does not automatically extend to any items identified as reprinted by permission of other publishers and copyright holders. Such items must be excluded unless separate permissions are obtained. It will be the responsibility of the user to identify such materials and obtain the permissions.

The publications of the National Council of Teachers of Mathematics present a variety of viewpoints. The views expressed or implied in this publication, unless otherwise noted, should not be interpreted as official positions of the Council.

Printed in the United States of America

Contents

Preface ix

Mentoring New Teachers: A Position of the National Council of
Teachers of Mathematics xi

Introduction 1
        Linda M. Fulmore, Mathematics and Equity Education Consultant, Cave Creek Arizona

Section 1: Why Mentoring Is Important 3

    The Blur of Preservice Teaching 4
        Mike Arney, Marshall and Swift/Boeckh, Atlanta, Georgia
        Eileen Fernández, Montclair State University, Upper Montclair, New Jersey

    What Are the Benefits of Mentoring? 5
        Hosin Shirvani, University of Texas—Pan American, Edinburg, Texas

    Mentoring Mathematics Teachers in the Twenty-first Century 6
        Shonda Lemons-Smith, Georgia State University, Atlanta, Georgia

    Mentoring Teachers in Standards-Based Mathematics Education: A Visual Framework 6
        Eula Ewing Monroe, Brigham Young University, Provo, Utah

Section 2: Who a Mentor Is 11

    Will You Be My Mentor? 12
        Gwen Carnes, Emporia State University, Emporia, Kansas

    Five Essential Responsibilities of an Effective Mentor 12
        Jeremy Winters, Middle Tennessee State University, Murfreesboro, Tennessee
        Jason D. Johnson, Middle Tennessee State University, Murfreesboro, Tennessee

    To My Once and Forever Master Mentor 13
        William D. Jamski, Indiana University Southeast, New Albany, Indiana

    Who Should Be a Mentor? 14
        Andrzej Sokolowski, Cypress Ridge High School, Houston, Texas

Section 3: What a Mentor Does 15

    The Five-Star Mentor: A First-Class Guide to Teaching 16
        Emily Peterek, University of Florida, Gainesville, Florida

    Mentoring a Preservice Mathematics Teacher 17
        Debra I. Johanning, University of Toledo, Toledo, Ohio
        Rebecca M. Schneider, University of Toledo, Toledo, Ohio

    Observing and Conferencing with a Focus 18
        Kathleen M. Clark, Florida State University, Tallahassee, Florida

Contents

What I Learned from My Mentor: Supporting Beginning Mathematics Teachers 20
Lynn Liao Hodge, University of Tennessee, Knoxville, Tennessee

Adding to the Repertoire: How Mentors View Their Roles 21
Nancy O'Rode, California State University—Northridge, Northridge, California
Nancy Terman, University of California—Santa Barbara, Santa Barbara, California

Putting the "Teacher" in Teacher Preparation 22
Peter Sheppard, University of Louisiana at Lafayette, Lafayette, Louisiana

Mentor-Teachers' Role in Connecting Preservice Mathematics Teachers' Learning with Real Classroom Teaching Settings 24
Shuhua An, California State University, Long Beach, California

Mentoring at Any Age Is Important 25
Ann M. Perry, St. Joseph's Academy, St. Louis, Missouri

Mentoring Mindsets for the Mathematics Teacher 26
Jason D. Johnson, Middle Tennessee State University, Murfreesboro, Tennessee
Michaele F. Chappell, Middle Tennessee State University, Murfreesboro, Tennessee

Effective Mentoring of Preservice Teachers 27
Hosin Shirvani, University of Texas—Pan American, Edinburg, Texas

"Aiding and Abetting" Teachers of Mathematics 29
Thomasenia Lott Adams, University of Florida, Gainesville, Florida

"I Don't Need to Have All the Answers" 29
Nancy O'Rode, California State University—Northridge, Northridge, California

So What Do You Want from Me? 30
Mary Belisle, Greely Middle School, Cumberland, Maine

My Mentors: Qualities That Made Them Special 30
Ann M. Perry, St. Joseph's Academy, St. Louis, Missouri

The Mentor as a "Fellow Worker" 31
Len Sparrow, Curtin University of Technology, Perth, Australia
Sandra Frid, Curtin University of Technology, Perth, Australia

Section 4: Tools for Mentors 33

Developing Effective Mentoring Skills for Mathematics Coaches 34
Phyllis Whitin, Wayne State University, Detroit, Michigan
David Whitin, Wayne State University, Detroit, Michigan

A Checklist for Scheduling Observations 35
Eileen Fernández, Montclair State University, Upper Montclair, New Jersey

Observing a New Teacher 36
Jason D. Johnson, Middle Tennessee State University, Murfreesboro, Tennessee

Using Videotaping and Stimulated Recall to Reflect on Teaching 37
Patricia A. Emmons, Lyme, Connecticut

Technology as a Communication Tool 38
Marshall Lassak, Eastern Illinois University, Charleston, Illinois

Talking about Teaching: A Strategy for Engaging Teachers in Conversations about Their Practice 39
    Margaret Smith, University of Pittsburgh, Gibsonia, Pennsylvania

A Guide for Reflecting on Mathematics Lessons with Beginning Teachers 41
    Nancy O'Rode, California State University—Northridge, Northridge, California
    Hillary Hertzog, California State University—Northridge, Northridge, California

Promoting Equity in the Mathematics Classroom 42
    Nancy Terman, University of California—Santa Barbara, Santa Barbara, California
    Nancy O'Rode, California State University—Northridge, Northridge, California
    Maria Guzman, Oxnard High School, Oxnard, California

Section 5: Ideas for Mentoring Programs 45

Forming a Cadre of Mathematics Mentors 46
    Alfinio Flores, University of Delaware, Newark, Delaware
    Cheryl A. Thomas, Arizona State University, Tempe, Arizona

Mentoring for High-Quality Instruction Using Adult Learning Theory: Lessons from Research and Practice 46
    Thomas J. Starmack, Bloomsburg University, Bloomsburg, Pennsylvania

Essential Components of a Novice Teacher Induction Program 48
    Patricia A. Williams, Sam Houston State University, Huntsville, Texas
    Sylvia R. Taube, Sam Houston State University, Huntsville, Texas
    Margaret A. Hammer, Sam Houston State University, Huntsville, Texas

Learning Never Ends: Meeting Mentors' Professional Development Needs 49
    Heather A. Martindill, Mid-continent Research for Education and Learning, Denver, Colorado

Virtual Mentoring in the Secondary Mathematics Teacher Education Program 52
    Jane Murphy Wilburne, Penn State Harrisburg, Middletown, Pennsylvania

Section 6: Lessons Learned 55

Challenges and Suggestions for Cross-Cultural Mentors 56
    Fatma Aslan-Tutak, University of Florida, Gainesville Florida
    Adem Ekmekci, University of Texas at Austin, Austin, Texas

Learning from a Novice Mentor's Mistakes 56
    Keith R. Leatham, Brigham Young University, Provo, Utah

Mentoring Bilingual Mathematics Teachers 57
    M. Alejandra Sorto, Texas State University, San Marcos, Texas

Building an Open Relationship: A Mentoring Vignette from Both Perspectives 58
    Michael E. Matthews, University of Nebraska, Omaha, Nebraska
    Nicole I. Guarino, University of Iowa, Iowa City, Iowa

Preface

> [S]eldom, if ever, do we ask the "who" question—who is the self that teaches? How does the quality of my selfhood form—or deform—the way I relate to my students, my subject, my colleague, my world? How can educational institutions sustain and deepen the selfhood from which good teaching comes?
>
> —Parker J. Palmer
> *The Courage to Teach*

All too often we hear anecdotes of teachers' leaving the field because they are overwhelmed by the demands of teaching. Perhaps a teacher education program has prepared a novice teacher with the necessary mathematics content knowledge, a foundation in pedagogy, some classroom discipline techniques, and hands-on experience in the classroom. Maybe a more experienced teacher is struggling to keep abreast of the constant barrage of changes in the field or within his or her own building. Even within the supportive structure of a university teacher preparation program, in-service teachers may feel weighed down by all the demands placed on them. Regardless whether one is a beginning, experienced, or preservice teacher, one can become overwhelmed by all that is required to merely survive let alone flourish as a mathematics teacher.

In the quote above, Parker Palmer challenges us to ask ourselves how we might help colleagues on their continuous journey to better their teaching. Do we leave our colleagues to flounder as they navigate all the complexities of what it means to teach mathematics? Mentoring is the answer to Palmer's question of how we might "sustain and deepen the selfhood from which good teaching comes." Mentoring can provide the support and encouragement not only to survive the demands and challenges of teaching but also to thrive and develop as professionals who are dedicated to the teaching of mathematics.

In 2004, NCTM published a series of publications titled *Empowering the Beginning Teacher of Mathematics*. Realizing that a gap existed in providing similar support specifically for mentors of mathematics teachers, NCTM's Educational Materials Committee issued a call for manuscripts that would provide the basis of practical "how to" advice for individuals who participate in formal or informal mentor training or serve in the capacity of instructional coach, peer coach, lead teacher, collaborative peer, department chair, administrator, critical friend, team leader, university supervisor, or department or grade-level colleague.

The original intent of the call was to create grade-level publications mirroring the framework of the beginning teacher books. However, when the editorial panel met to review the numerous submissions, the advice for mentors and mentoring programs was not so much differentiated by grade level but rather, was distinguished by the level of teaching experience of the teacher being mentored. The result is separate publications on the mentoring of beginning mathematics teachers, experienced mathematics teachers, and preservice mathematics teachers. Although some mentoring advice is specific to each group, other mentoring advice transcends any amount of teaching experience.

The intent of the editorial panel was to create publications on the mentoring of mathematics teachers that would be informative and practical resources that are easy to reference to address the reader's specific needs. We hope we have provided some useful ideas as well as challenged the reader to think differently about what it may mean to be a mentor.

REFERENCE

Palmer, Parker J. *The Courage to Teach: Exploring the Inner Landscape of a Teacher's Life.* San Francisco, Calif.: Jossey-Bass, 1998.

Mentoring New Teachers
A Position of the National Council of Teachers of Mathematics

States, provinces, school districts, and colleges and universities share responsibility for the continuing professional support of beginning teachers by providing them with a structured program of induction and mentoring. These programs should include opportunities for further development of mathematics content, pedagogy, and classroom management strategies.

> The retention of new teachers continues to be a concern in both the United States and Canada. Statistics show that nearly half of the new teachers in the United States leave the profession in their first five years of teaching, and Canadian and U.S. attrition rates are both around 30 percent for teachers in their first three years. These high rates of attrition contribute to the overall shortage of high-quality mathematics teachers, particularly at the middle school and high school levels. This attrition is especially alarming in the United States, where it is predicted that more than 2 million new teachers will be needed in the coming decade.

In far too many schools, new mathematics teachers receive challenging teaching assignments for which they are unprepared. These teachers, some of whom do not have strong backgrounds in mathematics content, are often isolated from professional involvement with colleagues. Frequently, they receive little content-specific professional development to support them in meeting the challenges that they face. As a result, their students may not be afforded the learning opportunities and quality instruction that the Council advocates as essential preparation for high-functioning adults in the workplace and everyday life.

Recommendations

States, provinces, school districts, and colleges and universities should provide professional development for new teachers by creating partnerships between experienced and novice teachers. These partnerships should ensure a strong focus on mathematics content knowledge, pedagogical knowledge, and knowledge of *Principles and Standards for School Mathematics* (NCTM 2000) and its application to the classroom. Education agencies should establish mentoring programs for new teachers and provide funding for the programs and the training of mentors. In making teaching assignments, district and school-based administrators should consider the additional demands on beginning teachers and their mentors alike. Teachers who have been identified as mentors should receive significant and consistent training, as well as appropriate remuneration or release time for their services. Finally, beginning teachers need and deserve a strong, structured program of induction, which includes mentoring, to ensure their success and increase the likelihood that they will stay in teaching, growing steadily in professional expertise and finding lifelong satisfaction in a career of continued service to mathematics education.

(September 2007)

Introduction
Linda M. Fulmore

Teaching is a rewarding, exciting, and challenging profession and commitment. Some would say it is complex because of the relationships that exist between teacher and student. Whereas teachers may focus primarily on content and pedagogy, students may focus on cognition and personal needs. Sometimes those two worlds of differing beliefs and attitudes about mathematics and learning come together with the goal of classroom experiences that lead to outcomes of high achievement and retention in the mathematics pipeline.

Today's preservice teachers are composed of both those whose initial college career goals included teaching and those who enter the profession mid-career. Each has differing educational experiences, preparation, and wide-ranging expectations and reasons for choosing teaching as a career. Thus the mentor and the preservice teacher each perceives the rewards, challenges, and complexities of teaching through a different lens. Mentors must be good listeners and perceptive of the needs of preservice teachers. Regardless of academic preparation and prior career experiences, all mentors reinforce the goal of high-quality instruction and classroom experiences for every student, every day.

Mentors who are knowledgeable, informed, and supportive have the potential to have a significant impact on the beliefs and practices of aspiring teachers. In a career field that experiences a high attrition rate, mentors can make a difference in attracting and retaining high-quality teachers. The primary responsibility of the mentor of a preservice teacher is to help the individual bridge the gap between the theoretical knowledge of teaching and learning mathematics with the reality of working with students in a school setting.

This volume outlines ideas and strategies to begin the process. Section 1, "Why Mentoring Is Important," lays the foundation for the need to mentor mathematics teachers. Sections 2, "Who a Mentor Is," and Section 3, "What a Mentor Does," discuss characteristics of mentors and suggest a wide variety of roles that mentors of preservice mathematics teachers might fill. Section 4, "Tools for Mentors," offers advice and suggestions to help the mentor support the preservice teacher, specifically proposing a number of helpful strategies for structuring conversations and observations. Section 5, "Ideas for Mentoring Programs," discusses existing programs and presents a variety of models to consider in developing a mentoring program. Finally, Section 6, "Lessons Learned," shares mentors' reflections and invaluable insights gained in their mentoring experiences.

This volume has a plethora of valuable information for mentors of preservice teachers. These practical ideas and strategies can supplement what you already do, or they can assist you in initiating a more formal mentoring program for preservice teachers. Just as you recognize that support and resources benefit preservice teachers, we hope that this publication will support and benefit your work as a mentor. Best wishes in your goal of mentoring the next generation of mathematics teachers.

Section 1: Why Mentoring Is Important

MENTORING preservice teachers might seem redundant to some. After all, the purpose of mentoring is to support the growth and development of teachers currently in the classroom, and a university teacher education program focuses on the development of future teachers. So why would one need to consider mentoring for preservice teachers?

Although the mentor of a preservice teacher might be someone at the university level, we typically use the term to refer to the cooperating teacher who teaches, coaches, and guides the future teacher about best practice in and out of the classroom. Few resources exist that outline expectations for the cooperating teacher working with a preservice teacher, beyond information that he or she may receive from the partnering university or college. Some may have had the opportunity to participate in training about expectations and how best to foster an enriching and productive relationship between the cooperating teacher and the preservice teacher. This volume attempts to synthesize the wisdom of cooperating teachers and bring it to the fingertips of all those involved in mentoring the preservice mathematics teacher.

Much of what we know about mentoring is that this special relationship should be founded on trust that is nonevaluative in nature. Although the university professor certainly gives support while instructing on best practice for the teaching and learning of mathematics, ultimately the professor must assign a grade to the preservice teacher, making this relationship evaluative. In addition to being a content and pedagogical expert, the mentor of the future teacher also offers the welcoming smile of a friend and a comforting shoulder to lean on.

Whether you are working with preservice teachers for the first time or have done so for countless times, the articles in this section give perspectives on why it is important to provide high-quality mentoring to an aspiring teacher.

❝ Those having torches will pass them on to others. **❞**

—*Plato*

The Blur of Preservice Teaching

This article contains an essay titled "The Blur" that was written for a larger study on the process of becoming a mathematics teacher. At the time of that study, the authors[1] were a mentor (Marianne) and a preservice teacher (Samuel). As part of that study, the authors worked together with nine other preservice teachers to create various written forums to document practice teaching from the preservice teacher's perspective. The result of this effort was a collection of essays on the preservice teaching experience written *by* preservice teachers *for* preservice teachers.

Samuel chose to write "The Blur" to document the complex and overwhelming feeling that is sometimes experienced when novices begin teaching. For those mentors whose preservice teachers are feeling overwhelmed, we reproduce the essay below with the hope that reading (and talking) about these feelings will better serve novice teachers as they learn to negotiate the multiple demands of the classroom.

> The first few weeks of student teaching, and to some extent the whole ten weeks, do not stand out clearly in my mind. Even at the time it was difficult to remember what happened in a particular class or on a particular day. When people would ask, "How is student teaching going?" I wouldn't know what to say. It was going, well or badly, I did not know. This is what I mean by the "blur" of student teaching. Incidents seemed to run into each other and get mixed up. Both individual classes and whole weeks were landscapes without contour, murky areas where you see only what is close by and what passes in an instant.
>
> The blur effect seems most commonly experienced by teachers in particular classes. So much happens (from the teacher's point of view!) in a single class that it is difficult to sort it all out on the spot: Where's the overhead? Oh, OK, don't trip on the cord. "Yes, I had a very nice weekend. And you?" Where's that transparency? There sure aren't many of them here yet, and it's time for class to start. What should I be doing now? I would start, but what's the point when I'll just have to repeat it for the ones still coming in. "All right, time to get started. Get out a piece of paper and try this exercise." Those numbers weren't very well chosen, were they? Why isn't Yvette doing anything? Who's that talking in the back? Is it about the problem? How much time do they need? How much time should I spend going over it? They won't all get it though, and they need to know this for the lesson today. What's happening as I write on the chalkboard? Can they hear me? Can they read my handwriting? Am I even doing this problem right? I wish I could see them. I'm sure Tamara and Mary are giggling about me right now. I should have worn better pants; these are really scruffy.
>
> The stream of consciousness becomes a raging torrent, or so it seems. If I had a moment without anything pressing and urgent to do, I felt sure something was wrong. Doing a math problem gives you certain things to think about: What is the best way to solve the problem? Can I generalize this method? What the heck do I do from here? Speaking in front of a group gives other things to think about: Am I being clear? How is my posture? What am I doing with my hands? Is my eye contact good? Am I getting my point across? Writing on a chalkboard or overhead introduces more thoughts: Can they read this? Turn off the projector before you erase. Will this make a screechy noise? Don't wipe chalk dust on your nose. Then there is the class to observe and keep in order: What is Francis doing with that ruler? Who was talking

---
1. All names in the article are pseudonyms.

over there? Are they getting this? How many are paying attention? Who should I call on? Why is Jackie the only one putting her hand up? Can I call on Charles without making him look stupid? Finally, the teacher has to think about the class period as a whole: How much time is left? Do they know their homework? When will I give back the quizzes? Should we spend more time on review, or are they ready to go on?

It's a lot to think about. Besides all that is going on inside the head of the teacher, she is supposed to figure out what is going on in twenty-five other peoples' at the same time. For me as a new teacher, there was sometimes so much to keep track of that I had no way of ordering, prioritizing, making sense of it all. Keeping track of the "big picture" is very hard when so many details feel so important. Of course, the "big picture" is "Are the students learning as well as possible in this class?" Getting an answer to that is already a challenge; figuring out what to do about a "no" is an even bigger one.

. . . I had a dream last night. Kerry, my main cooperating teacher, was introducing me to two university professors of mathematics I knew and admired. "They were watching you student teach," she said. "You know, through the window in the side of the classroom. They were very pleased. They know about the doubts you had, but they thought you were very dynamic and would make an excellent teacher."

—*Mike Arney and Eileen Fernández*

What Are the Benefits of Mentoring?

Research has shown that mentoring can have a positive impact on the preservice teaching experience and can be a powerful influence on the behavior and practices of mentees (Kuzmic 1994). Mentoring can increase employment retention of the novice teachers in cooperating schools (Odell 1986). Boyer (1999) found that attrition among new teachers who were mentored was significantly less than among nonmentored newcomers. Also, research has shown that teachers who were trained under the auspices of a mentor were more likely to use teaching strategies associated with effective instructional practices (Darling-Hammond 2000). In addition, mentoring programs produce mutual benefits for all participants. The mentee is given instructional and noninstructional supports, and the mentor has valuable opportunities for critical and constructive reflection on practice (Odell and Huling 2000).

— *Hosin Shirvani*

REFERENCES

Boyer, Katherine Lynn Williams. "A Qualitative Analysis of the Impact of Mentorships on New Special Educators' Decisions to Remain in the Field of Special Education." Fairfax, Va.: George Mason University, 1999. ERIC Document Reproduction Service no. ED 438643.

Darling-Hammond, Linda. "Solving the Dilemmas of Teacher Supply, Demand, and Standards: How We Can Ensure a Competent, Caring, and Qualified Teacher for Every Child." Kutztown, Pa.: National Commission on Teaching and America's Future, 2000.

Kuzmic, Jeff. "A Beginning Teacher's Search for Meaning: Teacher Socialization, Organizational Literacy, and Empowerment." *Teaching and Teacher Education* 10, no. 1 (January 1994): 15–27.

Odell, Sandra J. "Induction Support of New Teachers: A Functional Approach." *Journal of Teacher Education* 37, no. 1 (January–February 1986): 26–29.

Odell, Sandra J., and Leslie Huling, eds. *Quality Mentoring for Novice Teachers.* Washington, D.C., and Indianapolis, Ind.: Association of Teacher Educators and Kappa Delta Pi, 2000.

Mentoring Mathematics Teachers in the Twenty-first Century

With changing demographics in the United States, the need for effective mentoring of teachers, particularly mathematics teachers, has become increasingly important. Recent demographic data indicate that approximately 48.3 million children are enrolled in public schools in the United States in prekindergarten through twelfth grade. Of these students, 43 percent are members of a racial or ethnic minority group, 19 percent speak a language other than English at home, and 14 percent receive special education services (Rooney et al. 2006). Further, the U.S. Census Bureau (2006) reveals that about 18 percent of children under eighteen years old are living in poverty.

In the current era of high-stakes testing and accountability, a mentor must possess the knowledge, skills, and disposition necessary to provide effective mathematics instruction to all students, along with assistance to colleagues. Given the diverse, changing student demographics in the United States, the mentor and mentee's relationship must be grounded in the notion that "[a]ll students, regardless of their personal characteristics, backgrounds, or physical challenges, must have opportunities to study—and support to learn—mathematics" (NCTM 2000, p. 12). Further, an effective mentor is one who is knowledgeable about research in mathematics teaching and learning and best practices in the field.

The standards set forth by the Interstate New Teacher Assessment and Support Consortium (INTASC) provide a solid framework for mentors as they support, through deliberate scholarly planning and decision making, the development of mathematics teachers. Table 1.1 highlights how the INTASC standards are articulated with respect to mathematics teaching and learning.

For developing mathematics teachers, the mentor is a powerful and essential supporter and sponsor. The mentor functions as a bridge between teacher preparation, whether traditional or alternative, and effective membership in the mathematics teaching community. Whether mentoring a first-year teacher or someone with fifteen years of experience, the goal is the same: to provide "more and better mathematics for all students."

— *Shonda Lemons-Smith*

REFERENCES

National Council of Teachers of Mathematics (NCTM). *Principles and Standards for School Mathematics*. Reston, Va.: NCTM, 2000.

Rooney, Patrick, William Hussar, Michael Planty, Susan Choy, Gillian Hamden-Thompson, Stephan Provasnik, and Mary Ann Fox. *The Condition of Education 2006*. National Center for Education Statistics Publication 2006071. Washington, D.C.: National Center for Education Statistics, 2006.

U. S. Census Bureau. *Current Population Survey.* 2006 Annual Social and Economic Supplement. Washington, D.C.: U.S. Census Bureau, 2006.

Mentoring Teachers in Standards-Based Mathematics Education: A Visual Framework

Most of us who have mentored teachers in mathematics education are aware of an inherent and perplexing challenge: "Teachers, equipped with vivid images to guide their actions, are inclined to teach just as they were taught" (Ball, as cited in Wolodko, Willson, and Johnson [2003, p. 243]). In mentoring preservice and in-service teachers toward Standards-based practices, I have found that much of the work involves helping new teachers develop personal images, metaphors, and references that may be different from those to which they are accustomed.

Although I share my visual framework for Standards-based mathematics learning and pedagogy as a reference (see fig. 1.1), I urge teachers with whom I work to develop their own representations. The images they create seem to capture their personal views of mathematics education while revealing some of their concerns, as well.

Typically, a useful framework includes only enough detail to remind the teacher of major organizing ideas, concepts, and relationships; too much detail may detract from the purpose of the framework. The relationships among the various parts of the framework can be made explicit through placement, arrows, colors, font size and style, and so on. The real value of such an illustration, however, is not what is communicated to others but the ideas, concepts, and relationships the image evokes for the person who constructed it. The framework can be altered or completely reorganized as a result of additional learning or new insights.

Table 1.1

INTASC (Interstate New Teacher Assessment and Support Consortium) Standards Applied to Mathematics Teaching and Learning

Elements of Practice in INTASC Standards	Application to K–12 Mathematics Teaching and Learning
Standard 1: Knowledge of subject matter	Mathematics teachers should possess strong content knowledge in number and operations, algebra, geometry, measurement, and data analysis and probability.
Standard 2: Knowledge of human development and learning	• Mathematics teachers should understand developmental factors that affect learning and ensure that mathematics instruction takes into account multiple intelligences.
Standard 3: Adaptation of instruction to individual needs	Mathematics instruction should— • be differentiated to meet the needs of students at various cognitive levels, including those who are struggling and those identified as gifted or talented; • reflect principles of culturally relevant pedagogy or culturally responsive teaching; and • take into account students for whom English is a second language and students who have special needs.
Standard 4: Multiple instructional strategies	• Mathematics instruction should effectively use manipulatives, technology, and other tools for learning, along with a variety of modalities, such as student-centered learning, cooperative learning, and small-group and whole-group activities.
Standard 5: Classroom motivation and management	• Teachers should foster a mathematics classroom culture in which all student voices are equally valued and encouraged to participate in the learning process.
Standard 6: Communication skills	• Mathematics instruction should consist of high-level problem posing and questioning, as well as teacher-student and student-student mathematical discourse.
Standard 7: Instructional planning	Mathematics instruction should— • reflect high expectations and consist of high-level mathematical tasks; • build on students' informal mathematical experiences, prior knowledge, strengths, and interests; and • be aligned with local, state, and national mathematics standards.
Standard 8: Assessment of student learning	Mathematics instruction should afford students flexibility and multiple opportunities to demonstrate their mathematical understanding.
Standard 9: Professional commitment and responsibility	Mathematics teachers should strive to be reflective practitioners who critically evaluate their roles in promoting students' mathematics achievement.
Standard 10: Partnerships	Mathematics teachers should participate in professional learning communities, such as school-based, local, state, or national mathematics organizations.

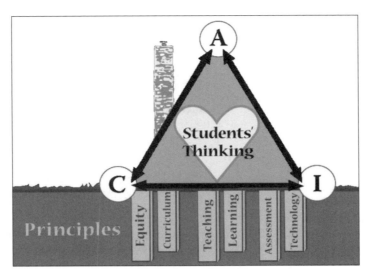

Fig. 1.1. Visual framework: My house of mathematics

For my visual framework, which continues to evolve, I have borrowed heavily from both my preparation and experiences as a teacher in various contexts, including the elementary school classroom situations in which I have served as a supervisor, professional development workshops, and university teacher education courses. The *Standards* documents of the National Council of Teachers of Mathematics (NCTM 1989, 1991, 1995, 2000), the writing and research of colleagues in the profession, and my experiences in supporting preservice and in-service teacher development have heavily influenced my model. The following paragraphs describe my framework, which I think of as a "house of mathematics."

The Foundation

Standards-based mathematics classrooms may look different from one another on the surface, but the knowledgeable visitor will see evidence of decision making guided by principles common to such classrooms. As NCTM notes, "The Principles for school mathematics reflect basic perspectives on which educators should base decisions that affect school mathematics. These Principles establish a foundation for school mathematics programs by considering the broad issues of equity, curriculum, teaching, learning, assessment, and technology" (NCTM 2000, p. 2).

The Structure

My house of mathematics is not showy, but it is strong. Notice that it resembles a triangle, considered to be the strongest of all basic shapes, with "corners" labeled with the letters C, I, and A. These letters stand for three major areas of teacher concern: curriculum and mathematics content, instructional pedagogy for teaching mathematics, and assessment. Arrows indicate the interaction of decisions made in each of these fields. I use the triangular model to illustrate the high level of interdependence among content, pedagogy, and assessment in making instructional decisions. I also use it to remind myself that although at times I may focus on mentoring teachers in one area, their success in promoting students' learning in their classrooms is dependent on a deep and connected knowledge base in all three areas.

The "Heart" of the Home

Over the years, I have become increasingly convinced of the need to place children's thinking at the center of our practice as teachers of mathematics. Any decision making regarding curriculum, instruction, and assessment should be based on knowledge of the mathematical thinking of our students and made with the goal of further promoting that thinking.

— *Eula Ewing Monroe*

REFERENCES

National Council of Teachers of Mathematics (NCTM). *Curriculum and Evaluation Standards for School Mathematics.* Reston, Va.: NCTM, 1989.

———. *Professional Standards for Teaching Mathematics.* Reston, Va.: NCTM, 1991.

———. *Assessment Standards for School Mathematics.* Reston, Va.: NCTM, 1995.

———. *Principles and Standards for School Mathematics.* Reston, Va.: NCTM, 2000.

Wolodko, Brenda L., Katherine Willson, and Richard E. Johnson. "Metaphors as a Vehicle for Exploring Preservice Teachers' Perceptions of Mathematics." *Teaching Children Mathematics* 10, no. 4 (December 2003): 224–29.

Section 2: Who a Mentor Is

THE MENTOR of a preservice teacher is someone who is aware of, and committed to, the time and energy required to mentor a preservice teacher. What makes mentoring a preservice teacher unique is the full-time commitment to that person. In other mentoring relationships, you and your mentee communicate occasionally, but in the end each of you is ultimately responsible for your respective classrooms. Mentoring a preservice teacher is a commitment to someone who has little to no experience in the classroom, and therefore it becomes the responsibility of the mentor to make real and tangible what was taught in educational methods classes.

Matching the aspiring teacher with an appropriate mentor is of utmost importance, as this relationship can have a long-lasting impact on the beliefs and practices of the preservice teacher. The previous section began to scratch the surface of the complexities of mentoring a teacher of mathematics. As we delve more deeply into the role of mentoring, we begin to explore the qualities and skills that a mentor might possess to have a beneficial and productive impact on preservice mathematics teachers.

" Mentor: someone whose hindsight can become your foresight. "

—*Unknown*

Will You Be My Mentor?

Will you be my mentor?
Will you be a helpful friend or a judge?
Will you appreciate my enthusiasm and be caught
    up in my excitement or ridicule my youth
    and inexperience?
Will you be my partner in new adventures and
    help me put my great ideas into practice?
Will you guide me with a gentle, experienced
    hand?

Can you travel back in your mind and remember
    what I am feeling and the kind of guidance
    I need?
Can you listen to my problems with an open heart
    and mind?
Can you help me learn from my mistakes?
Can you help me refine my teaching by sharing
    your experience?
Can you make time for me in your busy
    schedule? May I call you in the evening
    when I am worried about how to reach a
    troubled student or interact with an angry
    parent?

Do you really want to be my mentor? Am I a job
    or a professional responsibility?
Do you care about my success as you care about
    the success of your students?
Do you respect me? Do you view me as a peer?
Do you expect to learn and grow from our
    relationship?

Are you the right mentor for me?
Are you the person who will make a difference in
    my life, my teaching, and my career?
Are you the colleague who will help me become
    the best teacher I can be?
Are you the teacher I'll write about one day? "Let
    me tell you about my mentor, my friend."

The questions posed in this poem echo the reflections of novice teachers and mentors gathered during five year-round mentoring workshops in Oklahoma. These workshops[1] were intended to bring together novice teachers with a cadre of supporting mentors who were uninvolved in the first-year evaluation process. For the novice teachers, the workshops established a safety zone: They found a willing group of mentors eager to assist them and guide their progress. For the more experienced participants, the workshops fostered interactions with novice teachers that allowed them to redefine their roles as mentors. These experienced teachers found the role of mentor to be more than a professional obligation. Being a mentor requires the physical, mental, and emotional commitments and contributions of a good friend.

—Gwen Carnes

Five Essential Responsibilities of an Effective Mentor

Mentors play an essential role in the development of new teachers and, to be effective, must carry out certain activities in support of their protégés. The following list of five essential responsibilities of an effective mentor was gathered from discussions with mentor-teachers and protégés.

Providing a Solid Foundation

As a main contributor to the support structure for a new teacher, the mentor assists his or her protégé with day-to-day administrative tasks to allow the protégé to focus on teaching. Moreover, the mentor supports the protégé in handling discipline issues that might hinder effective teaching.

Sharing Ideas and Information as a Good Colleague

A good colleague might be defined as a coworker who takes on the other four responsibilities listed here. As a good colleague, a mentor keeps his or her protégé informed about school events or traditions, shares ideas, and co-plans. True collegiality benefits both the protégé

---

1. The workshops were funded through the Oklahoma Teacher Education Collaborative, a project of the National Science Foundation; the Oklahoma State Regents for Higher Education; and the University of Central Oklahoma.

and the mentor. The protégé gains the benefit of the mentor's insider knowledge of the school system in which they both work while the mentor learns new ideas from the protégé and may be inspired to vary classroom activities by the process of knowledge sharing.

Acting as the Protégé's Cheerleader

Beginning teachers need praise and encouragement. An effective mentor realizes when his or her protégé is struggling and tries to boost the protégé's morale. The mentor may work with the protégé during planning time or offer encouraging words over a cup of coffee or a meal after school.

Promoting Reflection on Practice

Good mentors encourage reflection on practice in their protégés. Having new teachers reflect on the events that take place in the classroom rather than tell them what went wrong or right helps them develop self-reliance and the skills to become better teachers. In this role, mentors need to strike a balance between providing support and micromanaging their protégés. Encouraging reflection helps new teachers think about and solve problems for themselves, not become mimics of their mentors.

Offering Constructive Criticism

Finally, effective mentors provide timely and relevant feedback. Protégés need encouragement and praise, but they also need to know where they can make improvements in their teaching. If the mentor has provided a firm foundation for the protégé, acted as a good colleague and cheerleader, and encouraged reflection on practice, then the protégé is more likely to be receptive to constructive criticism. If the mentor has not fulfilled these roles, the protégé may become defensive. Receiving feedback that enables learning and improvement is essential to the growth of the protégé. Mentors must learn to provide this feedback in a way that prompts acceptance rather than rejection.

—*Jeremy Winters and Jason D. Johnson*

To My Once and Forever Master Mentor

Dear Jim,

I was thinking the other day that it has been almost forty years since we first met! It seems like only yesterday that I was a twenty-one-year-old student-teacher of high school mathematics, as green as green can be, and you were a veteran of many years, classes, and students; the mathematics department chair then and for my six years at the school; and a professional interested in all the numerous "_CTM" organizations, whether G for Gary, I for Indiana, or N for National. You certainly knew a lot of mathematics, which you shared with our students—and me. When they (or I) did not understand a new idea, you always had "one more way" to show it. When I did not have a car to get to a "super" mathematics teacher meeting, you provided door-to-door service and back. When I forgot my coat for the annual individual school picture, you shared yours with me and another teacher so that we would "look professional." (Later readers, paging through that yearbook, will wonder if that coat was some kind of 1970s fad.) As a colleague, no matter how busy you were getting ready to teach a class or night school, you always had time to discuss students, the art of teaching students or teaching mathematics, my upcoming marriage, my children, or my plans to return to graduate school for a doctorate. In the dictionary, the word *mentor* should be—and in my mind and heart is—accompanied by your picture.

The years have passed quickly, and I thought that I would share with you some of my personal news: I am still married, I have a great family, and I got that graduate degree in mathematics education. In more than thirty years since I left the old school, I have taught more than 1500 secondary preservice educators, most just as green as I was, with more than 200 of them specializing in mathematics. Of course, during that time, I have also worked with their supervising teachers, and Jim, I think you would be proud to call this fine group of individuals colleagues. They fulfill many of the same roles with their protégés as you did with me.

When my mathematics student-teachers ask me what their student teaching and their supervising teachers or their first year of "real" teaching and their mentors will be like, I tell them that I hope their experiences will be as rewarding as mine was with a master mentor, you.

Always,
Bill

—*William D. Jamski*

- Analytical skills enabling the mentor to find meaningful solutions to the mentee's concerns
- Flexibility to offer a variety of teaching suggestions
- Passion for using new teaching techniques and delivering new content
- Knowledge and desire to monitor and assess the mentee's progress and identify areas for improvement

—*Andrzej Sokolowski*

Who Should Be a Mentor?

Anyone seeking to become a mentor should reflect on whether they have the interpersonal and professional attributes listed below.

Interpersonal Attributes

- Enthusiasm for the subject and the ability to inspire other teachers
- Belief in the mentee's ability to be successful as a teacher
- Willingness to encourage beginning teachers to try new instructional techniques and methods
- Ability to foster trust and respect
- Interest in listening to others, openness to suggestions, and the ability to be fair-minded and sensitive to others' needs

Professional Attributes

- Strong content knowledge
- Effectiveness as a teacher

Section 3: What a Mentor Does

THE first two sections of this book explored why mentoring preservice teachers of mathematics is important and examined characteristics and qualities sought after in someone who mentors. Yet many readers of this book may have skipped the first two sections and turned immediately to this one. Just as classroom teachers do, mentors search for ideas and suggestions on how to improve their craft. In this instance, the craft is mentoring.

Simply picking up this book demonstrates that you recognize the value of mentoring preservice mathematics teachers. Maybe you are already involved with mentoring at some level, and still you realize that you can, perhaps, do more in your role as a mentor. As you read through this section, you will find that a mentor's roles and responsibilities can be varied and far-reaching. The numerous articles that follow will undoubtedly support some existing beliefs you have about mentoring teachers of mathematics while also helping you think about mentoring in new ways.

> **“** Whoever ceases to be a student has never been a student. **”**
>
> —*Unknown*

The Five-Star Mentor: A First-Class Guide to Teaching

Five-star (five·star: *of first class or quality*[1])

The job of a mentor is to acclimate and empower a mentee with the goal of turning out an independent and successful worker or, in our relationship, a confident and effective teacher. I have been fortunate over the years to learn and grow under some brilliant mentors—influential individuals who were able to encourage but not push, support but not coddle, suggest but not criticize, and lead but not dictate in the context of mathematics education. With this background, I have reflected on my growth under the tutelage of those phenomenal guides in an effort to suggest what I believe to be the cornerstones of effective mentoring. These cornerstones are described in the following paragraphs to enable others to benefit from what I call the five stars of good mentoring.

Investment (in·vest: *to involve or engage*)

Both mentors and mentees must be "invested" in their relationship. The mentor, however, is primarily responsible for developing a shared sense of commitment. Initially, most mentors sketch out tentative plans for their mentees. Such proposals, which may include observation schedules, lesson ideas, and collaborative meeting times, should be discussed at the first meeting and revisited thereafter to ensure that the needs of the mentee are being met. Further, the mentee should have a voice in issues to be addressed. After all, he or she may be nervous about teaching, student relationships, or basic survival in the classroom; thus, the mentor may focus on these aspects of the job early in the relationship. Providing useful, timely support fosters a sense of investment in both parties.

Protection (pro·tect: *to cover or shield from injury, damage, or destruction*)

Mentees must feel safe in their new environment. As I remember, walking into a building full of twenty-year veteran teachers, experienced administrators, and impressionable youngsters can be quite intimidating. Mentors and others must make new teachers feel welcome by inviting them to participate in collaborations, lesson planning, and similar activities. Further, new teachers bring a sense of novelty and creativity to the job. If comfort and safety are established, they may be more willing to take chances in mathematics instruction. In my rookie year, I was fresh from a great teacher preparation program and full of ideas. Rather than sift through each lesson I had planned, my mentor, Noreen[2], encouraged me to try out my ideas: "You can always modify activities in the future," she said. "After all, if you never make a mistake, how do you learn?"

Consistency (con·sis·tent: *marked by harmony, regularity, or steady continuity*)

Being a mentor is difficult. After all, you have your own classroom, your own life, and your own well-being to worry about. The good news is that most mentees understand this reality. They do not expect their mentors to be in the classroom after every class period to answer questions, nor do they expect to receive baskets of engaging teaching materials every week. In fact, a mentee may be so wrapped up in acclimating to his or her new surroundings that excessive attention may be suffocating. The mentor should be available, however, if needed. Before school began, Noreen and I met a few times to discuss questions I had and to have a little bit of fun. She stopped by several times during my first week, offering to sit down and talk if I needed to but not pushing me to confide in her. Occasionally, she would send me an inspirational e-mail message or nice note or catch me in the copy room for a chat. Soon, I was off on my own, joining various committees and organizing assorted events. Her steady, reliable support, though, continued to help me immensely.

Longevity (lon·gev·i·ty: *long continuance*)

Investing in your mentee and offering initial protection may seem easy at first, but mentors should remember that these types of support must be lasting. The relationship is not sustained for just a few weeks or months but

---

1. All definitions are taken from the *Miriam-Webster Online Dictionary* (http://www.m-w.com).

2. A pseudonym.

until the mentee feels comfortable as a teacher. Even then, he or she may come to you with questions or suggestions or just to talk, and your support at these times must be unwavering. Of course, all teachers have busy weeks, and experiencing a few days of minimal communication does not mean that all is lost in the relationship. For mentees, just knowing their mentors are available for the long haul helps get them through the day. From September to May, Noreen observed my class; assigned me to observe others; and checked on me through e-mail, informal meetings, and notes. She backed off slowly, letting me get my feet wet, but somehow, I always knew she would be available for me. This relationship continued during my tenure at the school, and although years have passed and we have both moved on to other educational settings, we still correspond through e-mail. Our relationship has passed the test of longevity.

Extension (ex·ten·sion: *an enlargement in scope or operation*)

The relationship between mentor and mentee should also be extended beyond school. An occasional cup of coffee or dinner out may encourage both parties to open up and discuss important topics on their minds. Being physically removed from the school setting allows both mentors and mentees to "let loose" while still discussing school issues. Such occasions may also offer the opportunity for mentees to meet other school employees who may work in different departments or locations, broadening the support and social circle for new teachers. Mentees will greatly appreciate any efforts you make to help them relax and have some fun.

Indulging your mentee in luxurious, five-star treatment is sure to produce first-class results in a new teacher.

—*Emily Peterek*

Mentoring a Preservice Mathematics Teacher

Mentoring is an important strategy for supporting novices in learning to teach. Typically, however, descriptions of mentoring programs are focused on supporting teachers during their early years of teaching. Yet mentoring can and should begin even earlier, during a teacher's preservice preparation. An experienced teacher who takes a preservice candidate into his or her mathematics classroom also serves as a mentor. Because mentors support novices in learning to teach, being an expert mathematics teacher is not enough. Mentors must also acquire skill in developing professionals.

To support new mentors, we developed a graduate-level course for mathematics teachers hosting preservice teachers enrolled in a middle-grades methods class. Our work with new mentor-teachers highlights two important ways in which a mentor-teacher can support a preservice teacher. The host mentor-teachers realized that candidates particularly need help with translating plans into enacted lessons and with using assessment results to make informed instructional decisions. Although these ideas may seem like common sense for practicing teachers, our mentor-teachers identified them as important areas on which to focus support for preservice teachers engaged in initial experiences teaching mathematics to middle school students.

The Interaction between Planning and Teaching

Novice teachers often need support enacting their initial lesson plans. One mentor shared the following about her candidate:

> The written out lesson plan was fantastic. It was extremely thorough. Everything you wanted to see, I saw plenty far in advance. The issue the student teacher had was the scripting—she did not actually know what to do with it. She wasn't sure, and she was very unsure of herself in some regards. So every day we would sit down and walk through the next day's lesson. We basically taught the next-day lesson to each other in a condensed form so if we came across something, we could script it out and run into roadblocks and try to solve them before they happened.

Other mentors commented on the importance of supporting preservice candidates by providing feedback during the planning process. Preservice candidates do not have a history with the students in the classes in which they are practicing planning and teaching. Candidates need help understanding which concepts are easier for students and which concepts are more difficult for students to grasp. Preservice teachers also need support in interpreting students' reactions to their lessons. "You make your plans, but you have to be flexible enough to react to your students' grasp of the concepts." Timing is another area in which preservice candidates need support when they plan and teach a lesson. For example, looking over and discussing a preservice candidate's lessons is an opportunity to help the candidate realize how much time a task or activity may actually require to implement. Reflecting on a lesson after it is taught is also an important activity to help the preservice teacher process the outcomes of his or her enacted lesson plan.

Assessment as a Way to Guide Planning and Teaching

A second area of support that mentor-teachers identified is the important role that classroom assessment plays in learning to teach. Because novice candidates are new to planning and teaching, they often focus heavily on developing and implementing activities to assess students' progress. Learning how to use the information from assessments is an area in which candidates need support. Although assessing and evaluating middle school students' learning of mathematics is essential to ensure students' progress, mentor-teachers also realized that assessment of students' learning had clear benefits for their preservice candidate. One mentor wrote, "I think they [candidates] need to do some of their own assessment to see if what they intended actually happens, not just informal, but formal as well." Another mentor talked about embedding assessment into instruction as well as at the end of the unit. This mentor explained that doing so would serve two purposes—first, helping the mathematics students stay focused on the content, and second, providing feedback for the preservice candidate about how the students were progressing toward learning and understanding that mathematics content. In that sense, mentor-teachers are helping candidates recognize the connections among planning, teaching, and assessment.

Conclusion

Because the preservice candidates described were with their mentor-teachers for only two to three weeks, these mentors realized that the candidates still needed support in many areas. One mentor commented that "three weeks was not a lot of time to learn, but for my candidate it was enough time to learn a lot." We have described two areas highlighted by mentors as important and realistic for their novice preservice candidates who were just beginning to learn how to teach mathematics.

—*Debra I. Johanning and Rebecca M. Schneider*

Observing and Conferencing with a Focus

The university supervisor–preservice teaching intern relationship is one of the first formal supervisory relationships that occur in a teacher's career. University supervisors of preservice teaching interns are consistently called on to do many tasks simultaneously. Throughout a typical semester of supervising a preservice teacher, the university supervisor's tasks include conferencing with the intern and the cooperating teacher; observing the preservice teacher; completing assessments of the preservice teacher's performance; attending or coordinating a seminar for all the mathematics preservice teaching interns; and completing a variety of administrative forms—most of which focus on the preservice teacher's abilities as a new teacher. Additionally, the supervision tasks of the university supervisor may represent only one small facet of his or her professional work. So how can a university supervisor attend to just one pair of related tasks—the observation of, and follow-up evaluation conference with, the preservice teaching intern—so that both individuals involved in the experience grow in their practice as current and future mathematics teachers?

Mentoring Reflective Practice

One idea for focusing the observation and evaluation of an intern's preservice teaching is to select a focal point that both the supervising teacher and the intern can engage in equally. Such a concentration is not meant to deemphasize the other goals of the preservice teaching experience. Instead, a focus for the mutual investment of the university supervisor and the intern can offer coherence to a sometimes frenetic compilation of experiences.

During one experience as a university supervisor, Keri[1] and I decided to focus on her use of questioning practices in teaching mathematics. We spent sixteen weeks of her preservice teaching semester (eight weeks in high school "intensive" geometry; eight weeks in middle school honors geometry) examining the kinds of questions she asked and for what purposes she asked them. The role of questioning in teaching was emphasized in Keri's preservice education program, which was guided by the reforms posited by the National Council of Teachers of Mathematics. Despite an emphasis on such reform-minded teaching, however, "student teachers have been found to typically begin their classroom teaching experiences posing too many testing questions and hardly asking any provoking questions" (Ferreira and Presmeg 2004, p. 5). With this finding in mind, Keri and I set out to use my observations of her teaching and her vision of the role of questioning for mutual analysis of this aspect of her teaching.

The Experience

Throughout the internship, Keri and I decided to address the role of questioning in her teaching through three different activities. First, we decided to engage in dialogue journaling, which enabled us to carry on written conversations about Keri's use of questioning in her teaching over the entire semester (Garmon 2001) and in ways that were not limited by post-observation conferences. Second, we used the observation notes of Keri's questioning practices to guide our conversations reflecting on her use of questioning, her declared purpose for particular questions, and ways in which questioning influenced her instruction and students' learning. Third, we met on several occasions after her internship ended to examine trends in Keri's questioning practices.

In one dialogue entry, Keri identified several kinds of questions that she believed she asked while teaching, including (1) association questions to activate previously acquired knowledge; (2) questions to check for understanding; and (3) qualifying and review questions. To facilitate a discussion about Keri's actual questioning practices, we examined transcripts of three 90-minute observations and identified eighty-nine distinct questions. Keri categorized the majority (60 percent) of the questions as being for the purpose of activating or using previously acquired knowledge. Yet we wondered whether this focus was the optimal use of questioning for engaging students. Nicol (1998–1999) observed that the prospective teachers with whom she worked "struggle[d] with not only how they might ask students questions but also what they might ask and for what purpose" (p. 53). This difficulty also held true for Keri, who shared with me her struggles with *how* to ask questions. In one discussion about a question that Keri posed to the class when reviewing how to find the area of a regular polygon, Keri stated that she was unhappy with asking, "What's the area of the whole regular hexagon?" She decided that—

> The question is a little loaded. It's like, "okay, you're done; find it." Instead, I should have asked, "How can I go about finding the area of a regular hexagon? Can you try and describe the steps first before we actually calculate the area?" This is important because some students will be able to just do it and others will just feel like all they can do is sit there.

Although many of her questions were calling for students' procedural application of previously learned concepts, Keri realized the value of their use. Keri noted, "though my questions may be very directional and leading, by arriving at an actual answer the students will hopefully learn how to think logically and mathematically."

Questioning for Classroom Reform

Is value to be gained from engaging preservice teachers in questioning their questioning practices? Absolutely! Asking prospective and new teachers to identify the questions they think they pose for students and for what purposes—along with presenting them with a record of the questions they ask—engages the university supervisor and the student teaching intern in collaborative inquiry about how to translate their teacher preparation experience into a classroom culture that is rich in activities that "encourage their students' mathematical inquiry, understanding, and sense-making" (Nicol 1998–1999, p. 45).

—*Kathleen M. Clark*

---

1. A pseudonym.

REFERENCES

Ferreira, Rosa Antónia Tomás, and Norma C. Presmeg. "Classroom Questioning, Listening, and Responding: The Teaching Modes." Paper presented at the Tenth International Congress of Mathematics Education, Copenhagen, Denmark, July 2004. http://www.icme-10.dk.

Garmon, M. Arthur. "The Benefits of Dialogue Journals: What Prospective Teachers Say." *Teacher Education Quarterly* 28, no. 4 (Fall 2001): 37–50.

Nicol, Cynthia. "Learning to Teach Mathematics: Questioning, Listening, and Responding." *Educational Studies in Mathematics* 37, no. 1 (1998–1999): 45–66.

What I Learned from My Mentor: Supporting Beginning Mathematics Teachers

In my position, I mentor beginning mathematics teachers each year, and each year, I reflect on the qualities and responsibilities of a good mentor. As I think about effective mentoring, my thoughts often return to my mentor and my first year of teaching. Ellen (a pseudonym) was a seasoned veteran of twenty-five years, and she loved teaching. Through her mentoring, I started to make a contribution to my students as a mathematics teacher and learned about my identity as a teacher. Ellen was my "official" mentor, but much of the guidance she gave me was informal. Mentoring is no easy task, but Ellen made it seem effortless and comfortable, and our relationship was an open and supportive one. She inspired me to strive always to be better in the next lesson and to get to know my students as people, not just learners of mathematics. She always valued my ideas but challenged them in the spirit of debate that was indicative of her passion for teaching students mathematics.

I learned many things about teaching mathematics from Ellen, but my focus here is to identify her top three strategies for developing an effective mentoring relationship. I have tried to use these same strategies through various phases of my life, and they are invaluable to me as I interact with beginning teachers today. In my mind, the most important three strategies used by good mentors are to value beginning teachers' ideas, discuss the reasons behind instructional decisions, and provide an alternative lens through which the new teacher can view the classroom experience. I offer a glimpse of each strategy in the following paragraphs.

Value Beginning Teachers' Ideas

Ellen always asked my thoughts on activities, students, and teaching. Following one observation session, she began our formal meeting with the question "What did you think of the lesson? What do you think students learned about mathematics?" I talked at length about the aspects of the lesson I found useful and those I might change in the next iteration. She encouraged me to share my thinking, concerns, and questions with her, and she took time to consider my comments carefully. She treated me as a colleague, and both her gestures and comments emphasized the importance of my ideas. This seemingly simple act of respecting my thoughts had significant implications for my relationship with Ellen and my view of myself as a teacher. I felt that Ellen was invested in me and my teaching, and I felt competent in the classroom. She believed in me and conveyed a genuine interest in my thoughts on teaching, even though she was clearly the expert with valuable insights in most situations.

Discuss the Reasons behind Instructional Decisions

Ellen consistently initiated conversations with me about the reasoning behind instructional decisions. I learned a great deal from those conversations because they fostered insights into the decisions she made before, during, and after mathematics lessons. Her "suggestions" to me were not presented as lists of do's or don'ts. Rather, they surfaced through everyday conversations about instructional decision making. Through this practice, she also modeled reflection and an outlook that continuously examined instructional decisions in light of students' learning. Ellen emphasized the importance of understanding the reasons behind teachers' instructional practices, not only the effective or "best" practice itself. In retrospect, I believe that this understanding offers beginning teachers a toolbox of practices to work with and an understanding of why each one may be useful. They are then able to draw on and adapt practices to work in their specific classroom situations.

Provide an Alternative Lens

Often during our mentoring relationship, Ellen suggested different ways to look at classroom situations to give me insights that I had not yet discovered on my own. In this way, she provided an alternative lens through which to view events in the classroom. If I talked with her about a particular student's difficulties in learning a concept, she might offer a number of possible explanations for the situation. She might ask about the student's home life or about specifics of how the student solved particular problems. In doing so, she fostered the idea that many factors may affect classroom performance. Ellen rarely stated an answer absolutely; instead, she indicated potential responses. In this way, she helped me realize that teaching involves analysis and refinement and that it is rarely if ever perfect. In fact, her motto seemed to be that one of the most interesting aspects of teaching is that it can always be improved. Teachers can always do better in the next lesson.

Results

Through Ellen's support and expertise, I came to view reflection and change as natural parts of teaching. For her, mentoring was about open communication, trust, and mutual respect. She never pushed me to change, but she encouraged me to view change positively and to carefully and intellectually consider changes in my practice that would enhance students' learning and interest in mathematics. As we worked together, Ellen came to understand the teacher I wanted to become, and she respected my choices. She supported me in fulfilling my goals as a teacher instead of trying to make me a carbon copy of herself. Her task was not a simple one, but in my view, the strategies she used lie at the heart of effective, supportive, and thoughtful mentoring of beginning teachers.

—*Lynn Liao Hodge*

Adding to the Repertoire: How Mentors View Their Roles

During an intensive, three-year leadership development project, we asked thirty mentors each year how they viewed their mentoring roles. Every year, the mentors conducted twenty-one hours of seminars for beginning teachers, visited classrooms, taught model lessons, and coached beginning teachers; they were supported in these endeavors by participation in annual institutes and monthly leadership development meetings throughout the three-year period. As we analyzed the mentors' responses after each data-collection cycle, we found that the mentors played three general roles: resource, relationship builder, and change agent.

The Mentor as Resource

At the beginning of the project, mentors generally envisioned their role as providers of information and teaching strategies for beginning teachers. Mentors saw themselves as resources for ideas, materials, mathematical knowledge, and teaching tips. One mentor wrote, "I see myself as a resource to those teachers [whom] I will work with. My goal is to identify the individual needs of the beginning teachers. I want the teachers that I work with to feel positive about themselves and their math instruction. I hope my support will ease some of the frustration many of our new teachers feel about teaching math."

The Mentor as Relationship Builder

After the first year in the project, some mentors began to see the importance of building a trusting relationship with beginning teachers. Building a relationship meant listening, helping to create a safe environment for mathematics learning, and allowing teachers to communicate openly about their fears and successes in the classroom. Again, one mentor expressed a personal view of this role:

> During this year, I've gained a lot of confidence as a mentor. I have a growing understanding that my relationships are central and the most useful part of my program with the beginning teachers. As they begin to trust me, they ask questions more freely, and they begin telling me about strategies they tried and how they succeeded or failed. I've

done a great deal more listening than I expected, and I've also learned much from the teachers I initially set out to help.

The Mentor as Change Agent

We define a *change agent* as someone who takes responsibility for asking questions, introducing topics of equity and access, and bringing to the forefront issues that are sometimes ignored at the district, school, or classroom level. *Principles and Standards for School Mathematics* (NCTM 2000) states that equity is fundamental to high-quality mathematics education and, therefore, must be an essential element in developing teacher-leaders in mathematics. Over a three-year period, the project incorporated many opportunities for mentors to work collaboratively to define and explore equity issues. Generally, as mentors felt more confident in their roles, they brought up equity issues with beginning teachers more often. Some mentors in their second and third years wrote about their roles as change agents:

> I have always seen myself as a worker for justice, but up until a few years ago, I saw very little connection between equity and math. I now see increasing ways that equity affects student learning opportunities, especially in ways that are unspoken. I see my role expanding in leading students to see themselves as strong and capable, not limited by gender or color or class. And I am more willing to question, to speak up, and to stir up the comfortable acceptance of "the way things are."

> I see myself as an advocate for students who may not learn math in traditional ways. By helping to facilitate seminars for beginning teachers, I provide new teachers a chance to see and experience many different strategies for doing mathematics. We all experience and learn math differently. I view my role more as an agent of change.

Conclusion

Although we have documented three roles that mentors played in the mentoring process, we do not mean to suggest that mentors should shed one role to adopt another. Rather, mentors accumulate roles over time, continually adding to their repertoire.

We invite you to reflect on your responsibilities as a mentor and the various roles you have incorporated into your mentoring. We hope this project summary serves as a prompt for mentors to reflect on their multidimensional mentoring practice and see the importance of their work, not only in their own professional growth but also in supporting newcomers to become effective teachers of mathematics.

—*Nancy O'Rode and Nancy Terman*

REFERENCE

National Council of Teachers of Mathematics (NCTM). *Principles and Standards for School Mathematics.* Reston, Va.: NCTM, 2000.

Putting the "Teacher" in Teacher Preparation

Since fall 2003, State University[1] has divided the overall responsibility of preparing mathematics teachers among a "community of scholars" that includes content area faculty, curriculum and instruction faculty, and current secondary school teachers. This revamped program requires students seeking secondary education certification to obtain a major in mathematics and a concentration in education. This change also allows secondary school mentor-teachers to have a greater role in the development and preparation of secondary school mathematics teachers prior to the practice teaching semester. An examination of formal and informal evaluative comments from mentor-teachers in this program reveals that they view themselves as partners, recruiters, and role models.

Partners

Mentors in the program indicated that they feel a genuine partnership with the university owing in part to the consistent communication between the university and the mentors. Specifically, mentors participate in a week-long workshop that includes research talks and casual dialogue with faculty from the mathematics department, department of curriculum and instruction,

---

1. The university name is a pseudonym.

and the dean of the College of Education. This kind of communication continues during the academic year through monthly meetings and frequent e-mail messages. The fact that the mentors see themselves as partners in teacher preparation is a testament to the amicable relationship between the university and mentor-teachers. Both groups view themselves as interdependent; and they foster that interdependence through communication, compromise, and commitment, all essential components of effective university-school partnerships (Digby, Gartin, and Murdick 1993). In addition, the university regards mentor-teachers as valuable resources, and this esteem in turn has conceivably made the mentors feel sufficiently treasured and supported (Darling-Hammond 2003).

Recruiters

Realizing that the prospective teachers they supervise will earn a degree in mathematics and may be lured to other fields upon graduation, mentor-teachers indicated that recruiting is an essential part of their role. Mentors emphasize that they have made conscious attempts to persuade prospective teachers to consider teaching their primary career. Although the task of matching a career in a science or mathematics field against a career in teaching may be daunting, the mentors were convinced that teaching may prevail because of the intangible benefits in teaching that are unparalleled in other professions (Sheppard 2005). Notwithstanding the lure of other fields, mentors have been relatively successful in keeping talented mathematics students interested in mathematics education.

Role Models

A consensus among the mentors in the program is that they give prospective teachers a balanced representation of the profession. A widely accepted assertion is that teaching has numerous undesirable aspects; therefore, these mentors believe that it is also necessary for prospective teachers to experience the triumphs and accomplishments that help counteract the perceived negative aspects of teaching.

In addition, the responsibility of being a role model is enormously important to the development of prospective teachers, primarily because it helps them understand how to teach. Cochran-Smith (2005) points out that current teacher education places a "bright spotlight on subject matter knowledge," which consequently overshadows pedagogy and other areas related to education (p. 12). Similarly, the teacher-preparation program described in this article parallels the curriculum of mathematics majors with a few minor exceptions. Therefore, pedagogical courses have been placed on the proverbial back burner, and much of learning how to teach is designed to occur during field experiences with mentor-teachers (Sheppard 2005).

Conclusions

Although no authoritative definition of mentor-teachers' role in teacher preparation exists, universal tenets should certainly be included in the definition. Further, maximizing mentor-teachers' expertise and soliciting their input contributes to the development of their role. When the mentors' role includes being recruiters, role models, and partners, all stakeholders are the beneficiaries. Therefore, a concerted effort on the part of universities to recruit mathematics mentor-teachers who demonstrate high-quality teaching habits and possess admirable character traits may be advantageous. The result is likely to be invaluable. Conversely, aimlessly recruiting mathematics mentors who may not possess the skills mentioned is likely to result in unalterable negative effects on prospective teachers.

—*Peter Sheppard*

REFERENCES

Cochran-Smith, Marilyn. "The New Teacher Education: For Better or Worse?" *Educational Researcher* 34, no. 7 (October 2005): 3–17.

Darling-Hammond, Linda. "Keeping Good Teachers: Why It Matters, What Leaders Can Do." *Educational Leadership* 60, no. 8 (May 2003): 6–13.

Digby, Annette D., Barbara C. Gartin, and Nikki L. Murdick. "Developing Effective University and Public School Partnerships." *The Clearing House* 67, no. 1 (September–October 1993): 37–39.

Sheppard, Peter. "Redefining the Role of Science and Math Mentors." *Academic Exchange Quarterly* 9, no. 4 (Winter 2005): 229–33.

> Mentor-Teachers' Role in Connecting Preservice Mathematics Teachers' Learning with Real Classroom Teaching Settings

According to the National Research Council (Kilpatrick, Swafford, and Findell 2001), "teachers' knowledge is of value only if they can apply it to their teaching; it cannot be divorced from practice" (p. 379). Many recent studies have indicated the existence of a gap between knowing and applying in teaching (Luo and Li 1999; Stigler and Hiebert 1999; Wu and An 2006). In the current trend in teaching, nearly half of new teachers switch careers after five years. One of the major factors in attrition found by school leaders is that new teachers coming to them are not ready for the classroom. Mathematics educators must develop new programs to connect preservice teachers' learning with the realities of classroom teaching, thereby enhancing new teachers' knowledge and teaching ability.

The goal of this project was to connect preservice teachers' learning in K–8 mathematics methods courses with real classroom teaching settings at a local school. Six mentor-teachers were invited to participate in this project. With the support of the mentors, a weekly three-hour meeting consisting of four sessions was held: (1) preservice teachers attended one hour of instruction on knowledge of mathematics teaching; (2) preservice teachers observed a model lesson given by a mentor-teacher in a real K–5 classroom setting; (3) preservice teachers engaged in active discussion and reflection on their learning from the model lesson in their own classrooms; and (4) preservice teachers worked with children in their mentors' classrooms. Mentor-teachers played important roles in preservice teachers' learning in the following six areas.

Mentor-Teachers' Roles in Connecting Preservice Teachers' Learning

Mentors' support in providing model lessons

In this project, six mentors designed two model lessons each for a total of twelve model lessons that directly connected with the content and pedagogy of mathematics methods courses at the K–5 level in different content areas in the fall and spring. Each week, all preservice teachers in the mathematics methods course went to a mentor's classroom to observe an hour of real classroom teaching. Mentors also provided opportunities for the mentees to ask questions about the lessons immediately after their observations in the classrooms. This format allowed preservice teachers the opportunity for focused reflection on what they had just learned from the mentor's lesson.

Mentors' support in the transformation of preservice teachers' learning

After observing the mentor's teaching, the preservice teachers reconvened in their education classroom and discussed the strengths and weaknesses of the model lesson together with the group. Listening to others' thoughts and learning from one another enriched their individual reflections, which they wrote after the class. In their reflections they were asked to pose at least one question to the mentor regarding the observed lesson. The mentor read these reflections and provided feedback on each one, with detailed answers to the questions asked by preservice teachers.

Mentors' support in working with children in classrooms

Each week, the preservice teachers had a chance to work with children with learning difficulties in their mentors' classrooms for about forty-five minutes to one hour. The mentors identified the children who needed help and guided the preservice teachers on how to provide the best help to improve the children's learning.

Mentors' support in teaching mathematics in their classrooms

With multiple opportunities for observing a series of model lessons, the preservice teachers applied the knowledge and strategies they had learned by designing and teaching a minilesson in the mentors' classrooms. The mentors guided the preservice teachers on how to design a good lesson according to mathematics standards, students' needs, and their course requirements. The preservice teachers not only gained firsthand experience in teaching but also received timely feedback on their teaching from the mentors. This feedback helped them reflect on and redesign their lesson plans.

Mentors' support in conducting projects in their classrooms

In addition to observing a mentor's model lessons, the preservice teachers were required to conduct various projects in the field, such as student error analysis and assessment projects. In the error-analysis project, mentors helped the preservice teachers collect twenty students' errors, analyze error patterns, and design methods to correct the errors. In the assessment project, mentors helped the mentees analyze standards and textbooks, then design an assessment with twenty selective and constructive response items. Mentor-teachers allowed these assessments to be given to their students, and helped the preservice teachers analyze strengths and weakness of students' understanding at their grade level.

Impacts of Mentors from the Mentors' Perspectives

The principal of the school and the mentors highly praised this model. They believed that this model of mentoring not only helped preservice teachers and their students but also benefited the mentors in preparing their model lessons, reading preservice teachers' reflections about their lessons, and answering questions in those reflections. In addition, this process prompted mentors to reflect more on their instructions and glean insights from reading and answering questions in the preservice teachers' reflections. For example, one of the mentors said—

> In reviewing your students' reflections of their math course, I believe that the modeled lessons they are observing have proven to be extremely powerful and helpful in understanding various techniques and strategies demonstrated to teach math to all students. I have enjoyed modeling my math lessons for them, answering their well thought out questions and reading their response papers. I truly believe that as individuals entering the teaching profession, they are given a true sense of what being a teacher involves. By observing math lessons taught by experienced teachers in all grades, they are able to visualize true applications of the elementary students' learning. These college students have a great opportunity to gain first hand knowledge of an actual classroom setting, including any spontaneous events that may occur and how the teacher handles these situations. I believe that the students have enhanced their understanding of mathematical concepts from observation of an actual lesson, asking the teacher questions, and responding to the instruction in written form.
>
> As a "mentor" teacher, I have truly appreciated their great questions and written reflections. I consider myself fortunate to be a part of this great program and partnership ... and I am hopeful to see this program grow even further in the future.

The results of this project show that the mentors' role in supporting preservice teachers' learning is significant and that it has encouraged preservice teachers and classroom teachers to engage in a productive, inquiry-based process of learning and applying knowledge.

—*Shuhua An*

REFERENCES

Luo, S., and H. Li. *The Teachers' Ability.* Shangdong, China: The Shangdong Educational Publisher, 1999.

National Council of Teachers of Mathematics (NCTM). *Principles and Standards for School Mathematics.* Reston, Va.: NCTM, 2000.

Kilpatrick, Jeremy, Jane Swafford, and Bradford Findell, eds. *Adding It Up: Helping Children Learn Mathematics.* Washington, D.C.: National Academy Press, Mathematics Learning Study Committee, 2001.

Stigler, James W., and James Hiebert. *The Teaching Gap: Best Ideas from the World's Teachers for Improving Education in the Classroom.* New York: The Free Press, 1999.

Wu, Zhonghe, and Shuhua An. "Using Model-Strategy-Application to Develop Pre-service Teachers' Knowledge and Assessing Their Progress in Math Methods Courses." Paper presented at the 2006 AERA (American Educational Research Association) Annual Meeting, San Francisco, California, April 2006.

Mentoring at Any Age Is Important

For me, the top five benefits of having a mentor in my teaching career are as follows:

- My mentor provided a place to turn for all my questions.
- My mentor benefited from my enthusiasm.
- Having a mentor meant that I knew at least one person in the school building.

- Knowing I had a colleague to turn to for support boosted my confidence.
- My mentor kept our conversations confidential.

—*Ann M. Perry*

Mentoring Mindsets for the Mathematics Teacher

Experienced teachers understand that no one "right" way exists to teach mathematics, but having the "right" attitude is important for any teacher. Having the right attitude means maintaining a healthy mindset when teaching mathematics and sharing ideas about teaching with in-service and preservice colleagues. This article describes some essential aspects of a mentor's mindset when coaching in-service or preservice teachers on effective skills for mathematics instruction. In particular, we discuss the strategies of speaking mathematically during instruction, anticipating students' questions or misconceptions, and using a variety of examples to represent mathematical concepts.

Speaking Mathematically during Instruction

Experienced teachers generally realize that speaking mathematically during instruction means more than just telling students what to do. According to Bratina and Lipkin (2003), "teachers should model communications that are precise" (p. 11) in conveying mathematical ideas during instruction. Beginning teachers should make sure to use appropriate terminology and phraseology when teaching and should require the same behavior from their students. For example, in talking about a door, a new teacher may call the shape a *square* rather than a *rectangle;* if the teacher misuses terms frequently, students, too, may fall into the habit of using such terms incorrectly.

At times, new teachers may also omit necessary words in phrases or fail to clarify certain phraseology. For instance, a fraction is often defined as "a part of a whole." This definition may be appropriate for younger students, but it may lead to confusion in the upper elementary and middle grades when the definition of a fraction as "a ratio" is more appropriate. When discussing probability, a protégé might say, "The probability of landing on heads in a coin toss is one-half" without clarifying that this result is theoretical and that the probability may vary under experimental conditions.

Anticipating Students' Questions or Misconceptions

Anticipating students' questions or areas in which students may have difficulty can make instruction more effective, but foreseeing potential problematic areas in mathematics may not always be easy for beginning teachers. To address this weakness, the protégé and mentor may plan lessons collaboratively, thus helping the protégé organize more comprehensive lessons, prepare effective questions, and predict possible student errors. For example, a mentor-teacher explained to her protégé that second-year-algebra students often experience difficulty in translating between exponential and logarithmic forms of equations. Sharing this information with the protégé enabled the two to discuss a plan of action to reduce students' confusion.

For the mentor, helping the protégé anticipate students' questions assists in connecting the "big ideas" in the mathematics he or she teaches. For instance, experienced teachers often note that students perform better in learning the four basic arithmetic operations when they have a firm understanding of place value. Mentor-teachers can help their protégés discover ways to teach the basic operations that enhance students' understanding of this important concept.

Using a Variety of Cases to Represent Mathematical Concepts

Finally, experienced teachers often teach by using examples or introducing situations that require students to use inductive or deductive reasoning skills. Mentor-teachers know the importance of using a variety of examples or cases when applying inductive reasoning; however, new teachers may not be aware that, for example, they habitually use squares to signify quadrilaterals, which could lead students to associate only squares with quadrilaterals.

Ultimately, we believe that one of the most important tasks in teaching mathematics is to foster students' ability to use spatial visualization skills to construct their knowledge. Mentors may realize that students have trouble with graphs when the axes do not have a scale of 1, but new teachers may always use graphs with x and y scales of 1. When students are given examples to graph that call for large numbers, they may have difficulty plotting points because, on the paper they are given, they cannot fit a graph that extends that long. To address this misconception, new teachers should show examples of graphs that use a variety of x and y scales appropriate to the data given.

Conclusion

The mentor-teacher's outlook on the teaching and learning of mathematics is vital to the development of their protégé colleagues. The three mindsets emphasized in this article are essential for empowering mentor-teachers to develop supportive dispositions to help beginning teachers establish long-term careers in teaching mathematics.

—*Jason D. Johnson and Michaele F. Chappell*

REFERENCE

Bratina, Tuiren A., and Leonard J. Lipkin. "Watch Your Language! Recommendations to Help Students Communicate Mathematically." *Reading Improvement* 40, no. 1 (Spring 2003): 3–12.

Effective Mentoring of Preservice Teachers

The relationship between the mentor and mentee is complex and multidimensional. Anderson and Shannon (1988) define mentoring as a process involving one who is willing to be a role model, an encourager, and a counselor for the preservice teacher. Glenn (2006) states that the role of a mentor-teacher is to work with the preservice teacher as a colleague rather than as a boss. To accomplish this goal effectively, a balance of power needs to be struck between the mentor-teacher and the preservice teacher in a classroom. For example, the mentor-teacher should not be too reluctant to hand over some of the class activities to the preservice teacher, nor should the mentor allow the practicing teacher to take full classroom responsibility. The mentor-teacher's responsibilities include helping the preservice teacher learn effective teaching and instructional strategies, and to support him or her through sharing experiences and empathetic listening (Odell 1986). The mentor-teacher is seen to be a skillful practitioner, a coach or trainer, and a reflective friend (Brooks and Sikes 1997). Furthermore, the novice teacher expects that the mentor-teacher will provide moral support and guidance along with constructive feedback (Gray and Gray 1985).

A mentor should not be an evaluator of the preservice teacher (Friske and Combs 1986). The inclusion of an evaluation component in the mentor's role can cause preservice teachers to become untruthful to their mentors (Driscoll, Peterson, and Kauchak 1985). Odell (1986) found that novice teachers provide more honest reflection of their teaching when their mentors' role is to assist them rather than to assess them. Moreover, when mentor-teachers' responsibility is to guide rather than assess, preservice teachers have fewer problems getting constructive assistance from their mentor-teachers (Stroble and Cooper 1988). Mullen (2000) believes that the role of mentor should move away from that of teaching expert to that of colleague in a power-sharing environment. Therefore, Mullen believes in using such terms as *comentoring, mutual mentoring,* and *collaborative mentoring* rather than mentoring. Bona, Rinehart and Volbrecht (1995) found that placing the *co-* prefix before *mentoring* reconstructs the relationship in such a way that it is viewed as nonhierarchical. In the past, these relationships were based only on the knowledge of the mentor.

Common Characteristics of Effective Mentoring

Rowley (1999) and Anderson, Major, and Mitchell (1992) cite common characteristics of effective mentors:

- commitment to the profession of teaching;
- belief that mentoring is an effective tool to help novice teachers;
- exhibiting empathy toward their mentees by sharing their experiences with them;

- accepting their preservice teachers;
- being active, continuous learners in the field of education;
- providing instructional support;
- having a positive attitude; and
- providing honest, constructive feedback.

At the same time, in an effective mentoring program, practicing teachers should be willing to provide their mentors with critical feedback (Lane, Lacefield-Parachini, and Isken 2003). A crucial component for effective mentoring is the establishment of trust and honesty between the mentors and the mentees (Lane et al. 2003; Beck and Kosnik 2002; Taylor 2002; Koerner, Rust, and Baumgartner 2002). The partnership must be one in which both parties can be both respectfully critical and supportive. Research has shown that many preservice teachers are reluctant to ask for help when they need it. Therefore, effective mentors are those who have strong interpersonal communication skills and provide emotional support and guidance to their mentees even when they are hesitant to ask (Whitaker 2001). Mentors should have open communication with their mentees, and they should encourage their preservice teachers to ask for help and guidance whenever necessary (Maltas and McCarty-Clair 2006). Especially important is that mentors provide caring feedback and be able to share their own experiences with preservice teachers. This outcome can be accomplished by beginning feedback with a positive statement and ending with positive feedback (Lee et al. 2006; Yost, Sentner, and Forlenza-Bailey 2000). A well-designed and well-implemented mentoring program enhances the professional growth of both the mentees and the mentors (Ackley and Gall 1992; Manthei 1992).

Although the relationship between the mentor and mentee is complex, the rewards of a mentoring relationship that is able to develop a truly collaborative partnership benefits all involved, including students.

—*Hosin Shirvani*

REFERENCES

Ackley, Blaine, and M. D. Gall. "Skills, Strategies, and Outcomes of Successful Mentor Teachers." Paper presented at the Annual Meeting of the American Educational Research Association, San Francisco, California, April 1992.

Anderson, Debra J., Robert L. Major, and Richard R. Mitchell. *Teacher Supervision That Works: A Guide for University Supervisors.* New York: Praeger Publishers, 1992.

Anderson, Eugene, and Anne Lucasse Shannon. "Toward a Conceptualization of Mentoring." *Journal of Teacher Education* 39, no. 1 (January–February 1998): 38–42.

Beck, Clive, and Clare Kosnik. "Components of a Good Practicum: Student Teacher Perception." *Teacher Education Quarterly* 29, no. 2 (Spring 2002): 81–98.

Bona, Mary Jo, Jane Rinehart, and Rose Mary Volbrecht. "Show Me How to Do Like You: Co-mentoring as Feminist Pedagogy." *Feminist Teacher* 9, no. 3 (1995): 116–24.

Brooks, Val, and Patricia J. Sikes. *The Good Mentor Guide.* Bristol: Open University Press, 1997.

Driscoll, Amy, Kenneth Peterson, and Donald Kauchak. "Designing a Mentoring System for Beginning Teachers." *Journal of Staff Development* 6, no. 2 (1985): 108–17.

Friske, Joyce, and Martha Combs. "Teacher Induction Program: An Oklahoma Perspective. *Action in Teacher Education* 8, no. 2 (Summer1986): 67–74.

Glenn, Wendy J. "Model versus Mentor: Defining the Necessary Qualities of the Effective Cooperating Teacher." *Teacher Education Quarterly* 33, no. 1 (Winter 2006): 85–95.

Gray, William A., and Marilynne M. Gray. "Synthesis of Research on Mentoring Beginning Teachers." *Educational Leadership* 43, no. 3 (November 1985): 37–43.

Koerner, Mari, Frances O'Connell Rust, and Frances Baumgartner. "Exploring Roles in Student Teaching Placements." *Teacher Education Quarterly* 29, no. 2 (Spring 2002): 35–58.

Lane, Sheila, Nancy Lacefield-Parachini, and JoAnn Isken. "Developing Novice Teachers as Change Agents: Student Teacher Placement against the Grain." *Teacher Education Quarterly* 30, no. 2 (Spring 2003): 55–68

Lee, Suk-Hyang, Raschelle Theoharis, Michael Fitzpatrick, Kyeong-Hwa Kim, Jerald M. Liss, Tracey Nix-Williams, Deborah E. Griswold, and Chriss Walter-Thomas. "Create Effective Mentoring Relationships: Strategies for Mentor and Mentee Success." *Intervention in School and Clinic* 4, no. 4 (March 2006): 233–40.

Maltas, Carla Jo, and June McCarty-Clair. "Once a Student, Now a Mentor: Preparing to Be a Cooperative Teacher." *Music Educators Journal* 93, no. 2 (November 2006): 48–52.

Manthei, Judith. "The Mentor Teacher as Leaders: The Motives, Characteristics and Needs of Seventy-three Experienced Teachers Who Seek a New Leadership Role." Paper presented at the Annual Meeting of the American Educational Research Association, San Francisco, California, April 1992.

Mullen, Carol A. "Constructing Co-mentoring Partnerships: Walkways We Must Travel." *Theory into Practice* 39, no. 1 (Winter 2000): 4–11.

Odell, Sandra J. "Induction Support of New Teachers: A Functional Approach." *Journal of Teacher Education* 35, no. 1 (January–February 1986): 30–34.

Rowley, James B. "The Good Mentor." *Educational Leadership* 56, no. 8 (May 1999): 20–22.

Stroble, Elizabeth, and James M. Cooper. "Mentor Teachers: Coaches or Referees?" *Theory into Practice* 27, no. 3 (Summer 1988): 231–36.

Taylor, Rosemarye T. "Shaping the Culture of Learning Communities." *Principal Leadership* 3, no. 4 (December 2002): 42–45.

Whitaker, Susan D. "Supporting Beginning Special Education Teachers." *Focus on Exceptional Children* 34, no. 4 (December 2001): 1–18.

Yost, Deborah S., Sally M. Sentner, and Anna Forlenza-Bailey. "An Examination of the Construct of Critical Reflection: Implication for Teacher Education Programming in the 21st Century." *Journal of Teacher Education* 51, no. 1 (January–February 2000): 39–49.

"Aiding and Abetting" Teachers of Mathematics

The phrase *aiding and abetting* may conjure up less than noble associations; however, when the phrase is examined for closer meaning, one discovers that *to aid* means "to help or assist" and *to abet* means "to encourage or support." This article examines the phrase *aiding and abetting* in the context of education and, more specifically, the context of mentoring teachers of mathematics.

As a mentor to teachers of mathematics, I see my role as aiding and abetting new teachers in building confidence and developing content knowledge. For many of the teachers I mentor, our interaction takes place through e-mail correspondence or telephone conversations.

Teachers I mentor commonly ask, "What do I do when …?" Usually the situation that comes after "when" involves a student's misunderstanding of a concept or a procedure. For example, one new teacher recently asked, "What do I do when my students confuse area and perimeter?" To aid and abet in these instances, I often begin by trying to understand why the mentee needs to pose such a question. My goal is not necessarily to answer the question but to empower the mentee to construct his or her own appropriate answer. In this instance, I asked the mentee, "Are the students confusing the concepts of area and perimeter or the formulas for area and perimeter?" Sometimes when teachers focus solely on the formulas using the typical variables of l and w (area = $l \times w$ and perimeter = $2l + 2w$), students can become confused about the formulas. To address this problem, the teacher can engage students in learning activities designed to help them build their understanding of the concepts, even to the point of generalizing formulas in a way that does not use the traditional variables l and w. By helping the mentee "step inside the mathematics" of area and perimeter as concepts rather than formulas, I aimed to help her develop strategies that would prevent such confusion among learners.

As you can see, I am guilty of aiding and abetting the mentees assigned to me—assisting and supporting their work as teachers of mathematics. I acknowledge this example as evidence of my guilt, and I accept my life sentence of commitment to mathematics teacher education!

—*Thomasenia Lott Adams*

"I Don't Need to Have All the Answers"

One of the best discussions we had at our monthly mentors' meeting was the one that began with one mentor's realization "I don't need to have all the answers."

The mentor was relating the unease she felt in leading her group of new teachers, knowing that she did not have all the right answers to their questions. After several months, she realized that she did not need to have all the answers; what she needed was to trust her beginning teachers enough to know that one of them would have an idea to solve the problem under discussion.

She then began to work on building trust in the group of new teachers so that everyone felt free to ask—and answer—questions.

As soon as this mentor was finished speaking, the others applauded her idea. Now our group happily shares the burden of finding answers with their mentees.

—*Nancy O'Rode*

So What Do You Want from Me?

The first few weeks of school are always tiring for me. For many years, I thought that the source of my trouble was the transition from the lazy days of summer to the craziness of the school year. As I grew into a more experienced teacher, however, I came to understand that those weeks represented a different challenge. I spent the first few weeks of the school year getting to know my student population and figuring out what each of the students needed from me to be successful. Does this student need a pat on the back or a little push? What can I do to help this student be more successful?

In the same way, when I started working with teachers, I discovered that my credentials and experience mattered less than the guidance I could offer to meet the needs of these adult learners. Student-teachers and teachers in their first few years of practice are faced with significant challenges. Like students, they represent a variety of learners with a variety of needs. When I figured out how to meet some of those needs, I became both valuable to, and valued by, these beginning teachers. When they came to me with questions that I was able to answer, my knowledge and experience were appreciated, but if I tried to give unsolicited advice, my words were less welcome. Although my responsibility was to guide them, the teachers had to feel the need for guidance first.

New teachers, like students, require a caring individual who values their contributions as developing professionals. Their successes are as varied as their backgrounds and life challenges. My support for them was only as effective as my discernment of what they needed from me.

—*Mary Belisle*

My Mentors: Qualities That Made Them Special

Because of the influence of my mentors, I profess the love of mathematics education to all those I meet. I credit my mentors with the fact that I have spent more than ten years teaching mathematics to children and that I do not know any bounds when teaching mathematics. Both mentors should feel proud of sharing a special part of themselves with me and guiding me to become the educator that I am today.

Both my mentors helped develop my professional teaching career. One was assigned to me under a state program that required every first-year teacher to have a mentor. The other chose me as a protégé because we had similar teaching philosophies. Both took the role of serving as my mentor as a serious responsibility. From my experience with these two mentors, I believe that the five attributes described below make a mentor relationship successful.

- Trust is a cornerstone of the relationship between mentor and mentee. Because I trusted both of my mentors, I told them about my classroom adventures willingly and without fear of punishment or reprimand. Beginning and, often, veteran teachers make mistakes in the heat of the classroom moment that we quickly regret. A mentor must be trusted to maintain confidentiality about such "lessons learned."

- Both of my mentors also exhibited positive attitudes. If I believed my lesson plans had not lived up to my expectations, they helped me implement activities and approaches that worked for me. In other words, they found the silver lining in my cloud of inexperience. They provided frequent encouragement when I needed it—sometimes on a daily basis.

- Both of my mentors had the knowledge and experience to offer suggestions for improvement in all aspects of classroom management, from discipline to techniques for teaching certain mathematics topics. The mentor should brainstorm with the beginning teacher to find ideas that work with his or her teaching style to ensure that the beginning teacher feels comfortable with the suggestions.

- Both of my mentors also exhibited an attribute that is difficult to find in today's world: the willingness to help. The mentoring relationship requires both this willingness and a significant time commitment from the mentor to the beginning teacher. This commitment may vary from an hour or two once a week to once a month and should be guided by the needs of the beginning teacher.
- The final attribute of a good mentor is belief in our profession. Many might think that any mathematics teacher automatically believes in the profession, but I would argue that this sense of commitment goes beyond the students. Good mentors realize that as teachers, we have a responsibility to provide guidance and support to newcomers in the field. I was lucky to find two mathematics teachers who believe in our profession and were willing to share that philosophy and their support with me.

—*Ann M. Perry*

The Mentor as a "Fellow Worker"

The following description of the qualities of successful mentors is based on the model of a "fellow worker" from the findings of Sparrow (2000). Important features of the fellow-worker model are choice, experimentation, and reflection. Principles based on these features are presented below.

The mentor-teacher should have experience working in the context of his or her mentee. The person acting as mentor must understand the context, expectations, and nuances of the environment in which his or her mentee works. For example, a secondary school teacher-mentor may have difficulty guiding or understanding the needs of a teacher who specializes in early childhood education.

The mentee should decide the problem or issue that will be the focus of attention. This principle empowers the mentee and removes the possibility of irrelevance or the imposition of the mentor's agenda. When the discussion is restricted to topics of interest and importance to the new teacher, any suggestions for action will be immediately applicable to the teacher's classroom.

Options for action should be generated through discussions between the mentor and the mentee. This principle embodies one of the main roles of the mentor, that is, to provide a range of possible approaches to address the problem or issue at hand. Each option should be analyzed in terms of its positive and negative aspects and how it fits into the general plan the mentee wishes to follow.

The mentee must decide which option will be implemented in the classroom or elsewhere to address difficulties that arise. This principle also empowers the mentee in that it enables a personal decision about what actions he or she can take given the constraints of the classroom, school, and his or her private life.

Mentors should encourage experimentation in the classroom to generate data and evidence for reflection. Having real data to analyze is essential to promoting reflection on classroom events. Such data are not derived from a book of theory but involve the mentee's students and teaching context.

Mentors should emphasize the importance of reflection on classroom practice and beliefs. The role of the mentor here is to highlight the positive outcomes of classroom experiments because these experiences are more likely to develop positive attitudes in the mentee. These experiences will then be repeated as the beginning teacher grows into the profession.

Professional development should be the "natural process of growth in which a teacher gradually acquires confidence, gains new perspectives, increases knowledge, discovers new methods, and takes on new roles" (Jaworski 1993, pp. 10–11). Teacher professional development through mentoring is based on the open sharing of ideas to give new teachers opportunities to reason about, and learn from, their own teaching experiences.

—*Len Sparrow and Sandra Frid*

REFERENCES

Jaworski, Barbara. "The Professional Development of Teachers: The Potential of Critical Reflection." *British Journal of Inservice Education* 19, no. 3 (1993): 37–42.

Sparrow, Robert Leonard. "The Professional Development of Beginning Teachers of Primary Mathematics." Ph.D. diss., Edith Cowan University, 2000.

Section 4: Tools for Mentors

LIKE lesson planning, being a mentor requires a good deal of time and preparation. The challenge for mentors of preservice teachers is to share their experiences and expertise and at the same time help novice teachers grow as professionals. Just as we want to empower our students to become independent learners, it is up to us as mentors of preservice teachers to help them become reflective practitioners. The most effective mentors are able to find the appropriate balance of supporting and guiding. Mentors offer support by sharing information, ideas, and advice. Mentors guide their protégés through coaching that involves listening and inquiry. A mentor is likely to have as many if not more questions about his or her responsibilities than the protégé has for the mentor.

The purpose of this book is to be a quick and accessible resource for mentors of preservice teachers. This section in particular furnishes numerous tools to consider, adapt, and use. Several of the articles are designed specifically for use with preservice teachers. For example, "A Guide for Reflecting on Mathematics Lessons with Beginning Teachers" is a user-ready tool that mentors might use to focus discussion about a particular lesson. Other articles offer mentors ideas and suggestions to help develop their own skills in supporting their protégés. "Talking about Teaching: A Strategy for Engaging Teachers in Conversations about Their Practice" suggests language for mentors to consider when engaging preservice teachers in reflections about their teaching.

Just as the instructional ideas, activities, and lessons that we borrow from colleagues are adapted to fit the specific needs, goals, and skills in our classrooms, these tools are intended to be modified to suit the relationship of individual mentors and protégés. Perhaps the best approach is to peruse the articles in this section and choose one or two that seem to match your needs as a mentor, adjusting the ideas where necessary to match your goals.

" No man is wise enough by himself. "

—*Titus Maccius Plautus*

Developing Effective Mentoring Skills for Mathematics Coaches

Mathematics coaches often face obstacles in their efforts to mentor classroom teachers. For example, when coaches lead a lesson, some teachers might view the activity as designed for the sole benefit of the students, not for themselves, or purely as an enrichment activity. As a result, they may remain uninvolved with the lesson, turning their attention to grading papers or other tasks. Other teachers are not confident in their teaching practices or knowledge of mathematics and resist involvement in coaching for that reason. These conditions are not conducive to effecting long-term change in the teaching and learning of mathematics. To address these concerns, a group of twenty mathematics coaches who served elementary and middle schools in a large urban district developed the following list of strategies for effective mentoring. These strategies focus on the collaborative planning of a mathematics lesson and on postlesson debriefing.

Suggestions for Planning a Lesson with a Teacher

- Discuss the context and climate of the classroom with the teacher to ensure that the lesson fits the teacher's style. For instance, does the teacher feel comfortable introducing a specific activity? Does the teacher use manipulatives often? Do the students work in groups on a regular basis? Have the students had much experience writing about mathematics? Answering these questions helps the teacher and coach work together to devise a plan that best fits the particular classroom.

- Invite the teacher to establish the focus of the lesson by asking, "What aspect of the students' learning do you want to know more about?" The teacher might want to focus on strategies to involve more students in the large-group lesson, to examine the ways in which students justify their reasoning or record their ideas, or to find out how English-language learners exhibit their understanding. Coaches may need to offer suggestions for a focus, but the teacher must make the final decision. In this way, the teacher develops a sense of ownership of the lesson. After the focus is decided, the coach may lead the lesson while the teacher gathers data that pertain to the question to be answered, or these roles may be reversed.

- Discuss with the teacher the reasons behind some of the mathematical decisions in the lesson. For instance, a coach might say, "After working with the students to find all possible ways to use factoring to count on to 10, I would like to challenge them with 6. I picked 6 because it, too, has four factors. I think that some students may be surprised that a number less than 10 has the same number of factors as 10. I want them to confront this result and suggest some theories about why it is true." Sharing this kind of thinking and reasoning helps the teacher understand the coach's instructional intentions.

- Capitalize on the teacher's knowledge of the students' abilities in determining the design of the lesson. How does the lesson meet the needs of all learners? Is the pace of the lesson appropriate? Which students need extra encouragement and support? This approach allows both the coach and the teacher to contribute mutually valuable insights that help ensure the success of the lesson.

Suggestions for a Postlesson Conference

- After the lesson has been conducted, hold a conference in which both the coach and the teacher share "three pluses and a wish." The three pluses are observations about aspects of the lesson that seemed successful. Either the coach or the teacher might refer to an interesting comment or personal connection made by a student, an intriguing question, an unexpected response, or another reaction from the class. The wish involves a plan for the future: How might this initial experience be extended? What changes should be made the next time the lesson is taught? What interesting features of the lesson should be supported in future lessons?

- Be honest and willing to discuss aspects of the lesson that might be handled differently next time. The coach in particular should point out moments during the lesson when he or she felt unsure. By sharing his or her own vulnerability, the coach builds a trusting relationship with the teacher. When coaches demonstrate a reflective stance and pose questions about their own teaching decisions, classroom teachers become more willing to share their questions and doubts.

- Discuss the focus of the lesson, perhaps using the three-pluses-and-a-wish format to frame the observations. If the focus was to challenge students to justify their reasoning, some topics to consider might include their abilities to give evidence, connect several mathematical ideas, note patterns, and make generalizations.
- Provide evidence to support observations, and encourage the teacher to do the same. For instance, if the teacher says, "The students seemed to have a good understanding of the pattern," the coach might ask, "What are some things the students said or wrote that helped show their understanding?" Encouraging each other to be specific helps the two members of the team analyze the experience comprehensively and better assess the students' competence.
- Discuss ways to link lessons over time to build students' understanding. Talk about how this particular lesson might be extended to enable the teacher to pursue related ideas in the future. The lesson should be seen not as an isolated event but as part of a larger plan for the class.
- Reflect on the collaboration process together. What aspects of the planning, teaching, and debriefing processes went well? How is the process of learning together as a coach and teacher similar to the kind of learning we want to foster in students?

Conclusion

These suggestions highlight the importance of a collaboration between teachers and coaches in which each participant shares resources and knowledge that contribute to the design of the lesson.

—*Phyllis Whitin and David Whitin*

A Checklist for Scheduling Observations

A wealth of information needs to be collected to schedule observations between mentors and preservice teachers. Sometimes forms are provided by the cooperating school or university for some or all of this information, but at other times, mentors and student-teachers need to make these arrangements on their own. This article presents a checklist of these specifics, along with some finer points that are not typically detailed even in forms that are provided.

The Checklist

Contact information

Have your mentee supply you with his or her home address, e-mail address, home telephone number, and cell telephone number. You should provide an e-mail address and work telephone number, and you may even consider providing a cell telephone number for use in the event of unexpected school closings (e.g., due to weather), illness, or schedule changes. Agree, ahead of time, on the best form of communication.

Directions

Have your preservice teacher provide you with directions both to and from his or her school. If your mentee has obtained the directions online, be sure to verify that he or she actually drives them (since online maps are not always accurate). Remember to provide a starting address.

School information

Include the school name, address (street address, town, and ZIP code) and telephone number.

Parking information

Include information about parking, and be sure to state whether extra time needs to be allowed for parking (particularly in urban areas). Also, directions from the parking area to the school may need to be supplied.

Unlocked doors

In this day and age of school security, be sure to include information about exactly how and where to get into the preservice teacher's school. Because most preservice teachers arrive early in the morning, they may be unaware of security measures taken during the day and of the access difficulties that these measures can present.

Cooperating teacher(s)

Include the name(s) of cooperating teacher(s) as well as courses the teacher(s) are responsible for.

Teaching schedule

Your preservice teacher must provide his or her teaching schedule. This information should include the names of courses, special student populations, room numbers (which sometimes additionally require building identifications), and dates when the preservice teacher will begin teaching the classes.

Special days

Include a schedule of school closures and altered schedules, particularly those related to spring breaks, religious holidays, state testing, special school events, and so on.

Observation dates

Some mentors and mentees prefer to schedule observation dates at the beginning of the practice teaching term. Some like to schedule them as the semester goes by. Typically, they set aside a period of time (e.g., one week) during which each of the first, second, third, and so on, observations will be conducted. Both you and the preservice teacher should be aware of these observation periods and should agree, ahead of time, on how observations will be scheduled.

Meeting place

Decide where, and at what time, you and the preservice teacher will meet in the school.

Announce your visit

Ask the preservice teacher to advise the main office of the day you will be visiting. Doing so will make your arrival and visit run more smoothly. Also, have your meeting place and cooperating teacher's name with you so that the office staff can more easily help you.

Set a deadline

Set a deadline for receipt of all this information. Remember that practice teaching remains one of the most overwhelming responsibilities of teacher preparation. Student-teachers easily, and justifiably, get caught up in the responsibilities of teaching, putting off administrative responsibilities. If a certain piece of information cannot be supplied by the deadline, the student-teacher should explain the reason (by the deadline) and inform the mentor of when it can be expected.

Concluding Remarks

You will likely want to amend this checklist to suit your particular needs as time goes by. Tending to this checklist will take the frustration out of scheduling observations, leaving your energy free for the real work of mentoring your preservice teacher.

—Eileen Fernández

Observing a New Teacher

Of course, mentor-teachers contribute to the professional development of preservice and in-service teachers, but how does a new mentor-teacher know what aspects of teaching to focus on when observing a protégé during instruction? From my experience working with mentor-teachers and protégés, I have noticed some essential elements that mentors should look for during protégé observation sessions. These elements are described in the brief paragraphs that follow. Highlighting these issues in a postobservation meeting will help guide the protégé in developing the skills to become an effective mathematics teacher. Mentors do not have to identify all these items during the course of observing one lesson. Rather, they may assess and discuss various elements over a series of several observation sessions.

Allowing for Adequate Transitions between Activities

The mentor should encourage the protégé to think about the transitions between various activities planned during instruction. For example, the teacher may need to shift the students' attention from group work to a whole-class discussion. The mentor may suggest strategies for making smooth transitions during the lesson.

Eliminating Distracting Verbal or Nonverbal Language

For the most part, protégés do not use distracting verbal or nonverbal language intentionally; however, the mentor may have to tactfully identify distracting behaviors to allow the protégé to control them. The protégé can work to eliminate unwanted behaviors after he or she becomes aware of them.

Fostering a Safe Classroom Environment

The classroom environment is also known as the *classroom culture.* All students should feel safe and free to ask and answer questions in this setting. The mentor should notify the protégé if the classroom culture is not suitable for supportive learning, and offer suggestions that will allow more openness among students.

Asking Questions That Require Students to Think

Protégés may not be aware that the questions they ask during instruction do not require students to use higher-order thinking skills. To remedy this problem, the mentor may wish to help the protégé plan lessons that include questions to promote higher-order thinking.

Emphasizing Student Talk over Teacher Talk

New teachers of mathematics sometimes believe that they must project the image of gatekeepers of knowledge; thus, they rely on lecturing instead of promoting student discourse. Mentors should help protégés structure the classroom to encourage student communication and give students opportunities to help other class members.

Focusing on the "Big Picture"

Many new teachers follow the textbook section by section while missing the main objective of a particular chapter. Mentors should remind their protégés to make connections with the major ideas in a particular unit during instruction. By planning together, mentors can help protégés see important connections across lessons and units.

Using Creative Examples

Using obvious or repetitive examples during instruction can limit students' understanding of a particular topic. For instance, habitually discussing squares when referring to quadrilaterals in general may cause students to identify only squares as quadrilaterals. Here again, the mentor may wish to plan lessons ahead of time with the protégé while encouraging the new teacher to include a variety of examples and counterexamples.

Keeping Track of the Time Frame of Events

Mentors may record start and end times for all activities or segments of a lesson to help protégés manage instruction time effectively.

Varying Instruction

Protégés seem to use the lecture format for instruction most often, bypassing other effective approaches. The mentor-teacher may encourage the protégé to explore other strategies during a lesson.

Using Correct Mathematical Terminology

New teachers may sometimes say *squares* when they mean *rectangles* or use the term *equation* when they mean *expression.* If teachers frequently misuse terminology, students may also fall into the same habit. The mentor should clarify misused terminology and encourage the protégé to be cautious when speaking mathematically.

—*Jason D. Johnson*

Using Videotaping and Stimulated Recall to Reflect on Teaching

Using videotaping and stimulated recall with a coach or mentor is one way to prompt teachers to reflect on and improve their instructional practices. The following paragraphs describe one lesson in which stimulated recall helped a teacher find ways to enhance her mathematics instruction.

The Lesson

In my role as an instructional coach, I videotaped Tony while she conducted a lesson about equivalent fractions in her classroom. During the lesson, students were involved in a paper-folding activity in which they folded a piece of paper into thirds, then colored two parts of the three to represent 2/3. Students were then asked to fold the paper in half and talk about the resulting fraction. When asked, "What fractional part is shaded?" several students called out, "four-sixths." The teacher asked whether 2/3 is equal to 4/6. A few students said yes, a few said no, and one adamantly stated that the two fractions were not equal. The teacher discussed more examples with the students, using folded paper as a model. By the end of the class period, most students were beginning to develop a conceptual understanding of equivalent fractions.

Postlesson Conference

The postlesson conference with Tony revealed some important findings about her teaching. When watching the videotape several days later, Tony revealed, "I never expected to have to do so many examples. I thought they would get it on the first or second example." When I pointed out to her that this lesson perfectly illustrated the technique of allowing students to construct their own knowledge using manipulatives, she responded, "I hate to say it, but we are under so much pressure most of the time that we don't have time to teach this way. I can't believe how quickly the time went. Where do I get the time to do this kind of activity when the homework in the book just asks them to fill in the numerators and denominators?" As she continued to watch herself on videotape, Tony became frustrated; she explained, "I got confused myself. I was hooked into trying to get them to understand the algorithm. I was trying to fit the activity with the algorithm, and I myself had trouble with it. I was speaking during the lesson, but my brain was saying, 'Wait a minute; we are confusing issues here.' They did not have any practice finding missing numerators or denominators, the approach used in the text."

Tony and I discussed some of her instructional decisions, paying attention to an activity that she started at the beginning of the class. "What was the purpose of that activity?" I asked. "Looking at the lesson now, I realize it had no connection whatsoever," she replied. "I could have left it out altogether." When I asked how omitting the initial activity would have helped the lesson, Tony said, "I guess it would have left me more time to do the paper folding." I then asked, "Can you think of how you could have allowed the students to explore the paper-folding activity, then led them to practice filling in missing numerators or missing denominators with equivalent fractions?"

Benefits of Videotaping and Stimulated Recall with a Coach or Mentor

The videotaping session was an enlightening experience for Tony. Her uninhibited ability to reflect honestly about her instruction with a coach was beneficial in increasing her effectiveness as a mathematics teacher. Discussions about her lesson clarified actions she needed to take to improve her instruction. She knew she wanted to allow her students to construct their own knowledge of the concept of equivalent fractions but struggled to align her ideas with the textbook.

Often, teachers work in isolation from their colleagues and are judged by supervisors and administrators who visit and observe only sporadically. When conducted effectively with a coach or mentor, videotaped lessons and postlesson conferences allow teachers to examine their instruction objectively, discuss their instructional decisions, talk about their teaching, and draw conclusions. This teacher clearly took advantage of the videotaped lesson and the presence of the coach as a sounding board to investigate ways to improve her mathematics instruction!

—*Patricia A. Emmons*

Technology as a Communication Tool

Future teachers in the alternative teaching certification program in mathematics at Eastern Illinois University complete a teaching internship lasting one academic year. During the internship, candidates in the program are employed as full-time teachers by a cooperating school and are responsible for teaching a full load of classes and handling all related teaching duties. Program participants fulfill these responsibilities without the benefit of a cooperating or supervising teacher. Instead, candidates are assigned a mentor by the cooperating school. Owing to a lack of funds and staff and for other reasons, most participants have little to no contact with their assigned mentors. As I work with these beginning teachers, I try to fill the mentor role online using—

- online discussions;
- message boards; and
- reflective self-assessments.

The use of online discussions and message boards allows program participants to read advice and post comments at times convenient for them. These tools also enable a timely response system with better flow and organization than e-mail, provide a running record of the discussions that have taken place among participants, and help the novice teachers track their progress. The data gathered online give evidence of growth and change during the participants' teaching experiences and become part of a database of information to draw on to help future candidates in the program.

I use reflective self-assessments as conversation starters that encourage candidates to think about issues related to teaching. These self-assessments are posted online so that candidates have time to organize their thoughts before posting responses. Below are sample reflections I have used to initiate discussions.

- How are you handling day-to-day issues related to teaching mathematics? Do you feel you have time for yourself?
- Describe how you came to your current classroom-management plan. How is the plan different from what you initially imagined?
- How are you handling homework issues? Do you assign homework? Do you grade it? If so, how? If not, why not?
- Describe a lesson that you felt was particularly successful. What made the lesson work? Did anything unexpected happen during the lesson? How much planning time did the lesson require?

Many program participants indicate that they appreciate this type of mentoring because it does not make them feel pressured or intimidated. They feel comfortable posting and responding to questions and having discussions online rather than face-to-face. More than one new teacher has said that the online mentoring system served as a consistent lifeline for help and advice that they could not get from their on-site mentors.

—*Marshall Lassak*

Talking about Teaching: A Strategy for Engaging Teachers in Conversations about Their Practice

One challenge faced by mentors is how to provide instructional support that enhances performance without judging or criticizing beginning teachers. Fostering a safe environment for conversations to take place is essential. At the same time, the mentor's use of nonthreatening language is important to sustain regular discussions about instruction. Consider, for example, the exchange that took place between Olivia[1], a preservice teacher completing a full-year internship in a middle school classroom, and her university supervisor, Nora. Nora was deliberate in her choice of words as she helped Olivia reflect on her teaching.

The Lesson

Olivia had just taught a lesson on finding the slope of a line represented in each of three forms: table, graph, and equation. As she launched the lesson, she drew the sketch shown in figure 4.1 on the board, reminding students that slope is the ratio of the rise to the run. As she drew the sketch, she correctly indicated that rise was the vertical change and run was the horizontal change. This distinction, however, was not clear in the completed drawing. As students worked in small groups to explore an assigned task, they exhibited some confusion about how to determine the slope from a table of values or a set of ordered pairs. Olivia patiently asked students questions about their work: What is the rise? What is the run? What is m? What is b? What is the equation? Still, some students did not understand what values to use in forming the ratio.

---

1. All names in the article are pseudonyms.

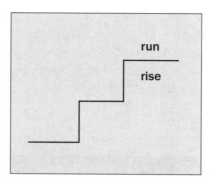

Fig. 4.1. Olivia's drawing

The Postlesson Conference

During the postlesson conference, Nora asked Olivia for her thoughts about the success of the lesson. Olivia noted that the students seemed to confuse rise and run, but she was not sure why. Nora pointed to a sketch she had made in her notes similar to the one that appears as figure 4.1 and told Olivia that she had *noticed* the diagram on the board at the beginning of class but *wondered* whether the labeling of the diagram might have been a source of the students' confusion. Nora's comments led to a conversation about how students might have misinterpreted the diagram as "run over rise." Nora and Olivia discussed other representations that might have helped students make sense of the concept of slope and the strategies for finding it.

Nora's deliberate use of *noticed* and *wondered* serves as an example of a nonthreatening way to talk about instruction. A mentor who *notices* something in a classroom draws on evidence from the observation session, for example, student work or field notes. The the mentor makes only factual statements, avoiding evaluation or personal preferences. Sometimes, the mentor may notice instances of good instructional practice, which should be reinforced with the beginning teacher to ensure that such practices are repeated. During the discussion with Olivia, Nora also noticed that Olivia had asked questions that made students think, as evidenced by their responses. For example, Olivia had asked, "If I give you a graph and you want to find the slope, what would you look for?" One student responded, "I would see if the slope was positive or negative by seeing which way it slanted," and another said, "I would make steps between the points and see how much it went up and over." Because questioning was something that Olivia had been working on, Nora wanted to make sure that she reinforced this aspect of the lesson.

In some instances, mentors may notice a practice that needs to be refined. The mentor should identify the practice and support his or her observations with evidence before engaging the beginning teacher in a discussion about how to improve his or her instruction. The use of the word *wonder* serves as a nonthreatening approach to launching this discussion. In this example, Nora used the word to engage Olivia in discourse that focused on reflection and problem solving related to representations of slope, an area of Olivia's practice that needed refinement. Engaging in such discussions may be the most difficult part of the mentoring process, but it is essential if growth is to occur.

The Noticing and Wondering Model

Mentors should structure a rigorous examination and analysis of practice by considering the beginning teacher's decisions and discussing both positive and negative results of instructional practice. The *noticing* and *wondering* model, shown in figure 4.2, is one way to promote such examinations of practice.

—*Margaret Smith*

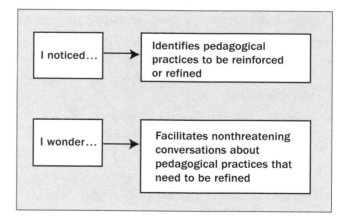

Fig. 4.2. Model for noticing and wondering

A Guide for Reflecting on Mathematics Lessons with Beginning Teachers

The beginning teacher you are mentoring has invited you to observe a mathematics lesson. Great! Now comes the hard part. How do you begin to discuss the lesson and stay focused on mathematics? Too often, issues of behavior or timing get in the way of discussing the mathematics in the lesson. One way to approach the discussion is to use the mathematics lesson reflection guide (see fig. 4.3), which incorporates the five Process Standards from *Principles and Standards for School Mathematics* (NCTM 2000). Asking about evidence of success concentrates the beginning teacher's reflections on tangible proof that students understand the mathematics being taught, and thinking about missed opportunities is a positive, less threatening approach to discussing needed improvements (Van Zoest 2004).

Using the guide in mentoring situations enables richer discussions between the mentor and the protégé that stay targeted on important mathematics.

—*Nancy O'Rode and Hillary Hertzog*

REFERENCES

National Council of Teachers of Mathematics (NCTM). *Principles and Standards for School Mathematics.* Reston, Va.: NCTM, 2000.

Van Zoest, Laura R. "Preparing for the Future: An Early Field Experience That Focuses on Students' Thinking." In *The Work of Mathematics Teacher Educators: Exchanging Ideas for Effective Practice,* AMTE Monograph 1, edited by Tad Watanabe and Denisse R. Thompson, pp. 119–34. San Diego, Calif.: Association of Mathematics Teacher Educators, 2004.

Mathematics Lesson Reflection Guide
Provide specific examples from the mathematics lesson for each of the boxes.

Process Standard	Evidence of Success	Missed Opportunities
Problem Solving Students actively engaged in problem solving.		
Reasoning and Proof Students made sense of the mathematics by explaining their reasoning and justifying their thinking.		
Communication Students expressed their mathematical thinking to others.		
Connections Students connected or applied mathematical ideas in appropriate contexts.		
Representation Students used multiple representations to model mathematical ideas.		

Fig. 4.3. Reflection guide based on NCTM Process Standards

Promoting Equity in the Mathematics Classroom

How do teachers promote equity in the mathematics classroom?

Why is reflecting on the issue of equity in the classroom important for teachers?

What are some characteristics of equitable mathematics classrooms?

In an effort to respond to these questions and help teachers think about equity issues in their practice, we developed an initial set of reflective questions. We further refined the questions through discussions with a group of teacher-leaders who were responsible for mentoring beginning teachers in mathematics.

We first identified four purposes for promoting equitable classroom practices, as follows:

- To provide access to mathematics for every student
- To develop the intellectual capacity of each student
- To encourage students to share their thinking and develop confidence as learners of mathematics
- To establish a classroom environment that is inclusive and respectful

The reflective questions in figure 4.4 (on the following page) are organized into groups corresponding to these four purposes. Teachers should use these questions as they plan lessons, reflect on their practice, or think deeply about the classroom environment. In our project, some teachers chose to focus on one or two questions at a time, using them to conduct informal research in their classrooms. Other project participants used the questions to promote discussion about important elements to consider in addressing equity in the mathematics classroom. The teacher-leaders found the questions to be useful in reflecting on their own practice and in promoting discussion with beginning teachers, but they agreed that the questions should not be used as a tool to evaluate protégés.

—Nancy Terman, Nancy O'Rode, and Maria Guzman

Equity in the Mathematics Classroom: Reflecting on Your Own Classroom

The following questions relate to equity in the mathematics classroom and have multiple uses. Use them in planning or reflecting on lessons and thinking about your practice or classroom environment. We suggest that you focus on only a few questions at a time, and keep in mind that this list is not comprehensive; you are encouraged to add your own questions.

Providing Access to Mathematics

Prior Knowledge

How does my lesson help students retrieve their prior knowledge about the topic?

Target Vocabulary

Are my students given opportunities to understand and use target vocabulary in ways that make sense to them?

Hands-on Activities

Does my lesson give all students the opportunity to work with manipulatives and other concrete learning tools?

Language

How does my lesson provide access for students at different reading and language levels and enable them to be fully engaged?

Learning Styles

Does my lesson consider the multiple and varied learning styles of students?

Developing Intellectual Capacity

Critical Thinking

How does my lesson promote higher-order problem-solving skills and intellectual rigor?

Mathematical Ideas

Do my students learn important and challenging mathematics through my teaching?

Mathematical Reasoning

Does my lesson allow students to question, make conjectures, and justify their mathematical thinking?

Expectations

How do I communicate and maintain high expectations for every student? Are all students challenged intellectually?

Encouraging Student Thinking and Discussion

Lesson Climate

How does my lesson help students develop confidence as mathematical learners and encourage individual thinking and risk taking?

Student Engagement

How does the lesson engage students? Are all students engaged? What are the opportunities in my lesson for students to work collaboratively, in small groups or pairs?

Student Participation

What efforts do I make to include all students in class discussion, especially those who are quiet or passive? Do I follow established rules for participation, such as enforcing a wait time to ensure that no one student dominates class time and teacher attention?

Developing a Culture of Respect and Inclusion

Meaningful Context

Does my lesson provide a meaningful context; that is, is it connected with students' background, culture, or experiences? Do students see what they are doing as important for their lives?

Culture of Inclusion

How is an overall culture of inclusion and expectation encouraged in my classroom? Do I explicitly espouse and practice inclusion? How does the physical environment of my classroom promote inclusion?

Climate of Respect

Do I promote and maintain a climate of respect for students' ideas, questions, and contributions in my classroom? How do I encourage students to bring their own ideas, thinking, and experiences to their learning?

Language and Behavior

Does my language and behavior clearly demonstrate sensitivity to issues of gender, race or ethnicity, special needs, English-language learning, culture, and socioeconomic status?

Challenging Bias and Stereotypes

Do I take opportunities to recognize and challenge stereotypes and biases that become evident among students? Do I analyze my interactions with students to check for biased language and stereotyping?

Fig. 4.4. Equity reflection questions

Section 5: Ideas for Mentoring Programs

WHETHER you are part of an existing preservice program, an adjunct university supervisor, or a teacher in a K–12 system, ideas and suggestions for how one might mentor and support preservice mathematics teachers are plentiful. Perhaps you are seeking ideas to supplement your current program, or maybe you are looking for suggestions and advice for others who work with student-teachers. Regardless, the ways to approach mentoring are as varied as the teaching and learning styles of our mentees and us.

At its core, mentoring is focused on the continual development of professionals. More specifically, in mentoring relationships for mathematics teachers, both the mentors and protégés continue to develop their content knowledge of mathematics, increase their pedagogical knowledge, and seek ways to blend the two to enhance students' learning—even when this relationship is between a neophyte and veteran teacher. Both parties benefit from the mentor-protégé relationship, and for this reason, mentoring programs can be designed to include components that target the professional development of mentors as well as the growth of preservice teachers.

Whether you are at the university working with future teachers or in the K–12 school system working with practicing teachers, the collection of articles that follow offer a plethora of ideas and suggestions, with topics ranging from criteria for creating mentoring programs to the professional development needs of mentors.

> **"** Mentors come in all ages and sizes, and with backgrounds as colorful and diverse as the signs in Times Square. **"**
>
> —*Heather A. Martindill*

Forming a Cadre of Mathematics Mentors

One of the goals of our school district was to give teachers the necessary professional development support to foster the teaching of mathematics for understanding. The strategy to make this profound change districtwide was to form a core group of mentor-teachers.

Project Structure

Mentors participated in two phases of the project. The first phase took place during the spring, and the second phase, during a summer session. In the spring, fifteen mentors took a graduate-level course in mathematics methods for credit, focusing on teaching in the middle grades. The course was offered at one of the schools in the district and met for three hours each week over fourteen weeks. The coursework gave special attention to topics that are important in the middle grades, such as number sense and place value; meanings of operations and operations with integers; fractions, rational numbers, and decimals; transitional concepts for algebra; ratio, proportion, and similarity; and functions and graphs.

The majority of the mentors participated in the summer session. The summer component consisted of twelve full-day workshops. During the morning, mentors observed lessons with students and activities modeled by the district expert. These lessons exemplified best practice.

The role of the teacher participants in these sessions was not passive. They invented word problems and interacted with students to provide learning opportunities. After the morning lessons, the mentors, curriculum facilitator, and students held debriefing sessions. Mentors analyzed elements of the lesson that were effective, improvements that could be made, and changes they could incorporate to make the lessons better fit their own styles and the needs of their students. In the afternoon, mentors formed small groups to discuss a daily focal question and other issues. Later, they engaged in activities using the same hands-on materials and addressing the same topics as the students but demanding a deeper level of understanding.

The support for mentor-teachers continued during the school year. Four-day sessions were conducted the next fall and spring following a similar format as the summer sessions. In addition, mentors met monthly. The mentors improved their own practice, provided leadership and support to other teachers in the district, and implemented standards-based curricula in the middle grades.

Positive Changes

Participants learned teaching methods that address national and state standards and that are based on research and best practice. These changes have been integrated into daily practice and have been sustained over time. Mentor-teachers have strengthened their understanding of content and the most effective methods for helping students learn mathematics. They are also more reflective about their own teaching. Further, participants now work with other teachers in their schools to disseminate their new knowledge. They have contributed to the district's overall improvement efforts as part of leadership teams. By building participants' profound understanding of the mathematics they teach, as well as expanding their pedagogical content knowledge, the mentor development program has contributed to sustained change in the district.

—*Alfinio Flores and Cheryl A. Thomas*

Mentoring for High-Quality Instruction Using Adult Learning Theory: Lessons from Research and Practice

Every good teacher plans for class by preparing introductions to ignite students' interest in the topic of the day, by developing activities to engage students with content, and by formulating essential questions to push students' thinking to a higher level. Mentors are good teachers who also have the knowledge and skills to assist their colleagues. Many mentors have received extensive training in instructional processes and pedagogy, that is, "the art and science of teaching [children]." But because mentors work with adult colleagues, they may benefit from a greater understanding of *andragogy*, "the art and science of helping adults learn" (Knowles 1980, p. 43).

Basic Assumptions of Andragogy

At the heart of andragogy are the following five major assumptions about adults as learners (Knowles et al. 1984; Knowles 1990):

1. The self-concept of adults is less dependent on the opinions of others and more internally motivated than that of children.
2. Adults have a wealth of experience that can enrich learning.
3. Adults' readiness to learn is closely related to skills necessary in their social roles.
4. Adults are more responsive to problem-centered learning than subject-centered learning.
5. Adults are motivated to learn by internal rather than external factors.

Various corporations, government organizations, and educational institutions apply these assumptions to their employee training. Researchers also agree that the assumptions of andragogy can be classified as principles of good practice (Merriam and Caffarella 1999).

As Elmore (2000) notes,

> ... the job of the administrative leaders is primarily about enhancing the skills and knowledge of people in the organization, creating a common culture of expectations around the use of those skills and knowledge, holding the various pieces of the organization together in a productive relationship with each other, and holding the individuals accountable for their contributions to the collective result. (p. 15)

Elmore's reference to administrators is applicable to the mentoring relationship. Mentors should focus on enhancing the skills and knowledge of their protégés while accounting for the specific characteristics and objectives of adult learners.

Suggestions for Practice

Applying the concepts of andragogy to collegial coaching is both easy and beneficial to those in the mentoring relationship. Below are some suggestions for putting the concepts of andragogy into practice, followed by brief explanations of the benefits gained.

- Maintain open communication by allowing the mentee to verbalize his or her achievements and problems while the mentor listens actively. By engaging in active listening, the mentor may be able to identify core problems. Mentors should restate problems they identify to clarify any misinterpretations. Working together to develop strategies to solve problems builds trust in the relationship. Mentors should also keep in mind that starting with smaller issues creates early success.

- Have an open discussion to set goals and expectations for the mentee, but avoid identifying too many goals. Having three primary goals in one school year is sufficient. Because mentors usually work with new teachers, they should keep in mind that their protégés may be overwhelmed by the process of acclimation to the system, the school, the students, and daily planning. The mentor should also be certain that the goals chosen are related to problems identified by the teacher. Additionally, mentors should incorporate goals for themselves. The mentor's goals can be simple, such as holding both mentor and mentee responsible for meeting expectations, providing resources for the teacher, or modeling deep, honest reflection.

- Have both parties agree to keep a journal with daily or weekly entries on conversations, thoughts, feelings, problems, strategies, and solutions related to teaching. The activity may have more depth and meaning if the mentor and mentee develop two or three essential questions to focus their writing, with the understanding that additional information may be added. Questions for the teacher might include these: What was my biggest struggle this week? Am I using my time effectively? Are my students engaged in my lessons, and how do I measure engagement? The mentor's questions might include the following: Do I actively listen to the concerns of my protégé? Do I allow my protégé to find his or her own solutions to problems? Semimonthly "journal swapping" allows both parties to gain a sense of the other person's thoughts and positions on important issues. Many people are guarded when speaking but tend to open up more in journal writing.

- Maintain the focus of the relationship by holding monthly meetings to reflect on initial goals and problems. If all the goals have been accomplished, begin the process again, identifying new objectives. Planning, implementation, and evaluation are elements of a collective, cyclic process. If one element is left out or the cycle is not repeated often enough, the whole experience is diminished.

Final Thoughts

Reflective practitioners are central to the process of mentoring. Serving as a mentor, coach, or administrator is an honor that allows one to learn more about pedagogy, get to know colleagues, and gain a greater understanding of the learning process. Through open communication and reflection, mentors not only assist their colleagues but also improve their own practice. Adult learning theory and practical experiences may serve as tools to be applied to reflection, learning, and growth for both mentors and protégés.

—*Thomas J. Starmack*

REFERENCES

Elmore, Richard F. *Building a New Structure for School Leadership.* Washington, D.C.: Albert Shanker Institute, 2000.

Knowles, Malcom S. *The Modern Practice of Adult Education: From Pedagogy to Andragogy.* 2d ed. New York: Cambridge Books, 1980.

———. *The Adult Learner: A Neglected Species.* 4th ed. Houston, Tex.: Gulf Publications Co., 1990.

Knowles, Malcom S., and Associates. *Andragogy in Action: Applying Modern Principles of Adult Learning.* The Jossey-Bass Higher Education Series. San Francisco: Jossey-Bass, 1984.

Merriam, Sharan B., and Rosemary S. Caffarella. *Learning in Adulthood: A Comprehensive Guide.* 2d ed. San Francisco: Jossey-Bass, 1999.

Essential Components of a Novice Teacher Induction Program

Leslie Huling and Virginia Resta served as the principal investigators for a $4.7 million grant from Houston Endowment, Incorporated, to develop a new teacher induction program (NTIP). The five essential components of the program provide a support system that goes beyond teaching strategies and curriculum development to address the realities faced by a new generation of novice mathematics teachers with unique needs and concerns. These five components are described in the following sections.

University Faculty Participation and Graduate-Level Classes

As part of the NTIP, participants completed six hours of tuition-free, graduate-level coursework. Mentees, university faculty members, and mentors met every other week to discuss topics of interest, such as classroom management, students with special needs, mathematics resource materials, and innovative instructional strategies that keep students motivated. The mentors and university faculty surveyed the first-year teachers to discover their needs and, at times, invited specialists to the meetings, for example, the district mathematics coordinator, who explained the curriculum. These three-hour graduate-level sessions helped build camaraderie among the new teachers in each school district. In some instances, the university professors taught classes off campus for the convenience of the mentees.

Collegial Atmosphere among Mentees

Not only did the mentees enjoy getting together every other week in class to share stories and exchange ideas, but they also kept in touch at other times. As part of the course requirements, these new teachers logged on to NiceNet, a free, Web-based learning environment. The friendly chat format of this site helped mentees feel at ease in asking questions.

Assignment of Individual Mentors

Most mentors in the program were retirees who were paid approximately $24,000 yearly to visit ten to twelve NTIP participants each semester. Weekly mentee visits ranged from quick stops to much longer meetings, depending on the new teacher's particular needs. Mentors assisted with a range of tasks, from modeling mathematics lessons to locating mathematics resource materials for English-language learners.

At times, the mentors became sounding boards while the new teachers vented their frustrations. As part of their role, mentors served as listeners, counselors, advisers, cheerleaders, confidants, and supporters. Bimonthly meetings and yearly mentor conferences helped the mentors gain new insight into their roles and allowed them time to share ideas. Mentors also had access to a NiceNet section where they could pose questions, seek advice, and tell success stories.

Cooperation of District and Campus Staff

Principals and counselors from each campus became role models to help novice teachers learn the habits, routines, and behaviors that reflected the beliefs, norms, and values of their particular schools. This support helped minimize conflicts between mentees' educational views and school expectations. Participating principals were informed about NTIP concepts ahead of time, demonstrated their support of the program, and agreed to abide by program expectations.

Access to Shared Teaching Resources

University professors, mentors, and mentees all shared teaching ideas and resources, such as books, journal articles, and Web pages, with other program participants. As part of the course requirements, the mentees created a *Bright Ideas* book that included their best lesson plans and proved to be an invaluable resource.

Impact of the Five Components

With these five components in place, novice teachers excelled and were retained in the teaching profession. The principal investigators currently track all NTIP teachers for five years, and the results to date indicate that more than 92 percent have remained in the profession. As one third-grade teacher, a mentee in the program, commented, "Keep helping those first-year teachers because it makes such a big difference."

—*Patricia A. Williams, Sylvia R. Taube, and Margaret A. Hammer*

Learning Never Ends: Meeting Mentors' Professional Development Needs

Mentors come in all ages and sizes and with backgrounds as colorful and diverse as the signs in Times Square. They have no single defining characteristic, but a description that comes close is *experienced*. Not surprisingly, districts often assign more experienced teachers to serve as mentors or coaches and ask these veterans to provide professional development support to novice teachers. Although mentors are experienced, they continue to have learning needs. A systematic approach to assessing and addressing mentors' learning needs can ensure that mentors and the teachers they guide are well equipped to face the challenges of mathematics instruction.

The "Taking Stock" Tool

The National Council of Teachers of Mathematics (NCTM) issued Standards for the Support and Development of Mathematics Teachers and Teaching in its *Professional Standards for Teaching Mathematics* (NCTM 1991). Standard 2 in this set, "Responsibilities of Schools and School Systems," states that school administrators and board members should "[provide] a support system for beginning and experienced teachers of mathematics to ensure that they grow professionally and are encouraged to remain in teaching" and should "[support] teachers in self-evaluation and in analyzing, evaluating, and improving their teaching with colleagues and supervisors" (p. 181). The form shown in figure 5.1, titled "Taking Stock of the Mentor's Professional Development Needs," allows mentors to express their objectives for professional growth and encourages districts and schools to assist in meeting those goals.

Although mentors may have a wealth of knowledge about instructional strategies and methods, they may need to hone their skills in sharing this knowledge with adult learners and helping teachers improve their practices. Again, the Taking Stock tool enables a mentor's supervisor to quickly assess the mentor's perceived knowledge and level of preparedness in three areas: content, instruction, and leadership. The tool consists of four sections:

Taking Stock of the Mentor's Professional Development Needs

I. ***To what extent do you feel prepared to help teachers ...***

	Not at all				To a great extent
1. use a variety of research-based instructional strategies to engage all (e.g., ELL, special education) students in important mathematics?	1	2	3	4	5
2. use assessment information to modify instruction for individual students and groups of students on an ongoing basis (i.e., throughout a unit, not just at the end of a unit)?	1	2	3	4	5
3. engage students in collaborative mathematics discourse?	1	2	3	4	5
4. communicate high expectations for all students through actions and words?	1	2	3	4	5
5. identify the "big ideas," key concepts, and knowledge and skills within the mathematics curriculum?	1	2	3	4	5
6. identify assessments that correlate to the conceptual understanding required and related knowledge and skills?	1	2	3	4	5
7. plan a standards-based lesson?	1	2	3	4	5
8. apply knowledge of the stages of change to adopt new instructional practices?	1	2	3	4	5
9. anticipate, identify, and resolve conflicts that are a result of the effort to improve mathematics teaching and learning?	1	2	3	4	5
10. recognize and celebrate improvements in their practices and their success in improving student achievement in mathematics?	1	2	3	4	5
11. establish and assess their own progress toward a goal for improving instruction and student learning?	1	2	3	4	5

II. ***To what extent do you feel prepared to ...***

	Not at all				To a great extent
1. systematically foster the sharing of ideas and successes between teachers in the school?	1	2	3	4	5
2. provide teachers with timely feedback that focuses specifically on the characteristics of high-quality mathematics instruction?	1	2	3	4	5

(Continued on next page)

Fig. 5.1. Tool for assessing mentors' needs and objectives: "Taking Stock of the Mentor's Professional Development Needs"

III. ***Using a scale of 1 to 5, with 1 being low, indicate your interest in learning more about each of the following topics:***

a. effective mathematics curriculum	1	2	3	4	5
b. high-quality mathematics instruction	1	2	3	4	5
c. standards-based instruction and lessons	1	2	3	4	5
d. helping teachers through the change process	1	2	3	4	5
e. providing feedback to teachers	1	2	3	4	5

IV. ***What problems do you anticipate will arise as you help teachers fully implement standards-based lessons and teaching? What might you need to learn more about in order to address these problems?***

What are two or three things you would like to learn next year to increase your ability to help teachers design and deliver high-quality, standards-based mathematics lessons?

(Denver, Colo.: Mid-continent Research for Education and Learning, 2007). Reprinted by permission of McRel.

Fig. 5.1. Tool for assessing mentors' needs and objectives: "Taking Stock of the Mentor's Professional Development Needs"—*Continued*

- Section I: Mentors assess their abilities to help teachers in the areas of instruction, curriculum, and leadership and the change process.
- Section II: Mentors respond to questions about their level of confidence in their coaching skills.
- Section III: Mentors rate their desire to learn more in the areas of instruction, curriculum, and leadership.
- Section IV: Mentors express additional needs or needs they cannot easily categorize.

Alignment of the Tool with Standards and Research

The Taking Stock tool is responsive to the Standards articulated in NCTM's *Professional Standards for Teaching Mathematics* (1991). For example, the questions on instruction are related to Standard 3, "Knowing Students as Learners of Mathematics," of the Standards for the Support and Development of Mathematics Teachers and Teaching. This Standard reads as follows:

> The ... continuing education of teachers of mathematics should provide multiple perspectives on students as learners of mathematics by developing teachers' knowledge of research on how students learn mathematics; the effects of students' age, abilities, interests, and experiences on learning mathematics; the influences of students' linguistic, ethnic, racial, and socioeconomic backgrounds and gender on learning mathematics; [and] ways to affirm and support full participation and continued study of mathematics by all students. (p. 144)

The questions on the Taking Stock form are also relevant to Standard 4 of the same set of standards, "Knowing Mathematical Pedagogy." Although mentors usually excel in mathematical pedagogy, this Standard divides this topic into important concepts that are essential to high-quality teaching. By addressing these concepts individually, the Standard helps mentors reflect on their own teaching and, thus, conceive ways to mentor others more effectively.

The Taking Stock tool is based in part on research findings by Weiss, Pasley, Smith, Banilower, and Heck (2003) indicating that only a small percent of mathematics lessons are of high quality; in fact, only 17 percent of elementary school lessons, 7 percent of middle school lessons, and 12 percent of high school lessons were found to be of high quality. Several of the questions related to curriculum and instruction on this form were formulated from the researchers' findings regarding the characteristics of high-quality mathematics lessons.

Other questions were derived from reports of the National Research Council titled *How Students Learn: History, Mathematics, and Science in the Classroom* (2005) and *How People Learn: Brain, Mind, Experience, and School* (2000) and from previous research on effective classroom instruction performed by Mid-continent Research for Education and Learning (McREL; see *Classroom Instruction That Works* [Marzano, Pickering, and Pollock 2001]). Finally, because mentors need to know not only the characteristics of high-quality lessons but also the methods they can use to help teachers change their classroom practices, the tool includes questions drawn from McREL's ongoing work on leadership and change.

Use of the Tool in School Districts

District-level coordinators can use the Taking Stock tool with mentors to determine the type and content of professional development opportunities their mentors need. Unsatisfactory results for a particular set of questions may highlight the need to address that area when investing in professional development. By using a tool that inventories staff needs, districts and schools can focus their professional development efforts more accurately and make better use of the vast experience their mentors have to share.

—*Heather A. Martindill*

REFERENCES

Marzano, Robert J., Debra J. Pickering, and Jane E. Pollock. *Classroom Instruction That Works: Research-Based Strategies for Increasing Student Achievement.* Alexandria, Va.: Association for Supervision and Curriculum Development, 2001.

Mid-continent Research for Education and Learning. "Taking Stock of the Mentor's Professional Development Needs." Denver, Colo.: Mid-continent Research for Education and Learning, 2007.

National Council of Teachers of Mathematics (NCTM). *Professional Standards for Teaching Mathematics.* Reston, Va.: NCTM, 1991.

National Research Council. *How People Learn: Brain, Mind, Experience, and School.* Washington, D.C.: National Academy Press, 2000.

———. *How Students Learn: History, Mathematics, and Science in the Classroom.* Washington, D.C.: National Academy Press, 2005.

Weiss, Iris R., Joan D. Pasley, P. Sean Smith, Eric R. Banilower, and Daniel J. Heck. *Looking inside the Classroom: A Study of K–12 Mathematics and Science Education in the United States.* Chapel Hill, N.C.: Horizon Research, 2003.

Virtual Mentoring in the Secondary Mathematics Teacher Education Program

Most secondary mathematics teacher education programs provide preservice teachers with opportunities to connect theory with practice through case studies, field-based experiences, laboratory schools, or even videotaped episodes of real-world classrooms. The quality of these practical experiences can make a difference in both the teachers' effectiveness and the likelihood of their remaining in the profession (Darling-Hammond and Bransford 2005). Involving entering freshmen and sophomores who are either planning or considering a career in secondary school mathematics teaching in practical classroom experiences can be difficult because of the number of general education courses they are required to complete prior to entering the teacher education program. However, these potential teachers need to be aware of the many joys, demands, and expectations of teaching as early in their teacher preparation program as possible. At our university, we implemented a Virtual Mentor program in which high-quality, exemplary middle and high school mathematics teachers serve as electronic pen pals to support and mentor undergraduate students, one-on-one, through the first three or four years of their college program. The goal of the virtual mentor program is to help acclimate candidates to the challenges and opportunities of teaching and to help them see the connections between their undergraduate courses and today's secondary school mathematics classrooms.

Preparing for Virtual Mentoring

Our mentors are recruited at professional development workshops and professional organization meetings or from among cooperating teachers, or are recommended by school district personnel. An initial e-mail message is sent to potential mentors to determine their interest in serving as virtual mentors. The e-mail message also states the purpose and benefits of the mentoring process. Once an undergraduate student is paired with a virtual mentor, the undergraduate student is encouraged to initiate the first contact with the virtual mentor to introduce herself or himself. All further correspondence between them is strictly confidential unless they are both willing to share information.

How involved the experienced teachers become as virtual mentors is up to them. We strongly emphasize that their role as a virtual mentor gives them the opportunity to make a difference in a future teacher's life. No paperwork is involved, and the mentorship is not expected to take up much of their time. They may serve as a mentor for one or more years, and with one or more undergraduate students. Also, if they have any questions throughout the process, they are encouraged to contact one of the mathematics education faculty members. A follow-up survey is typically given to all mentors and undergraduate students at the end of each year to receive feedback about how the mentoring program is working and whether they would suggest any modifications.

Benefits to the Virtual Mentor

The experienced teacher who voluntarily serves as a virtual mentor has the opportunity through e-mail correspondence to—

- interact with an undergraduate student in a secondary mathematics education program to share classroom experiences and expectations;
- engage in discussions regarding a particular classroom issue that arises;
- share interesting lessons and reactions from high school students with the undergraduate student;
- exchange information about mathematics content in the undergraduate program and its connection with the secondary school mathematics curriculum;
- support a college freshman by responding to questions and concerns they may have regarding a career in secondary mathematics education;
- invite the college student to his or her classroom to observe the class and meet students and colleagues;
- discuss new pedagogies;
- share various strategies that students have used to solve problems;
- acquaint the candidate with professional norms;
- have opportunities to reflect on his or her own teaching practices; and
- learn from an undergraduate student as the candidate shares what he or she is learning in mathematics and other college courses.

Benefits to the Undergraduate Student

The undergraduate student who voluntarily signs up to correspond with a virtual mentor has the opportunity through e-mail correspondence to—

- assess the complexities of teaching;
- acquire practical ideas and strategies;
- understand the workings of a classroom;
- understand issues facing educators in urban and rural school environments;
- provide real-world experiences; and
- connect content from undergraduate courses with secondary school curriculum or classroom issues.

Conclusion

Virtual mentoring is an excellent method of giving undergraduate students an opportunity to see the classroom from a teachers' perspective. And the earlier we can provide them with some types of classroom experiences, the better (Darling-Hammond and Bransford 2005). In our program, several undergraduate students took the opportunity to visit the mentor's classroom and even felt comfortable working with some of the students. The virtual mentors also noted that they found the opportunity to be especially rewarding. Answering some of the undergraduate students' questions and wonderings reminded them of their dreams and desires to become teachers. Overall, the benefits to both the mentors and the undergraduate students are sure to be a win-win experience.

—*Jane Murphy Wilburne*

REFERENCE

Darling-Hammond, Linda, and James Bransford, eds. *Preparing Teachers for a Changing World: What Teachers Should Learn and Be Able to Do.* San Francisco, Calif.: Jossey-Bass, 2005.

Section 6: Lessons Learned

THE MAIN focus of this book is on collegial and collaborative relationships between professionals who are dedicated to, and passionate about, the teaching and learning of mathematics. At one time, teaching was an endeavor that took place behind closed doors, with each teacher responsible for figuring out what was best for the students in his or her classroom. Teachers did not engage their colleagues in discussions about what worked or did not work in practice. Admittedly, some teachers have more opportunities for collaboration than others, but no one can deny the value of the learning that takes place between mentor and protégé—the protégé learning from the mentor and, often, the mentor learning from the protégé.

This last section, then, discusses "lessons learned" in the mentoring process. Who knows better than a teacher that life is about learning from our mistakes? Yet we can also learn from the mistakes of others, those who have traveled down the same road and are willing to share their insights. The authors in this section have performed precisely that service. Realizing that each of our experiences is unique, we cannot anticipate every situation or concern that will arise; however, if the advice offered here can improve the mentoring relationship even minimally, both the mentor and protégé will appreciate the experience all the more

> **“** Experience is a hard teacher because she gives the test first, the lesson afterwards. **”**
>
> —*Vernon Sanders Law*

Challenges and Suggestions for Cross-Cultural Mentors

During a course in mathematics methods taught by Fatma, a university instructor, preservice teachers engaged in a discussion about the use of children's songs in number-sense instruction. Fatma encouraged the preservice teachers to sing several songs, such as "One, Two, Buckle My Shoe" and "Ten in a Bed," and discuss how to use them in class with young students. Fatma commented, "I do not know these songs because I was raised in Turkey. I grew up with similar songs in my own language. Let's hear these songs from you." As she was walking around the classroom, one of her students softly said, "You are not an American if you don't know these songs!" The student's voice was just loud enough to ensure that Fatma picked up the comment but low enough that the other students did not hear it. The discussion continued, and Fatma acted as if she had not heard the student's remark. This insensitive act on the part of the student highlights the needs of those who mentor across cultures, particularly mentors who are members of a minority culture in a given context.

One of the first challenges international mentors usually face is a language difference. As nonnative speakers of English, we have sometimes borrowed teaching materials from colleagues who are native speakers, only to be told by students that the materials contain minor writing errors. The students are quick to blame these errors on the fact that English is a second language for us, but the same students ignore the errors when they are presented by our colleagues who are native speakers of English. Cross-cultural mentors should be aware of the possibility that some mentees may be more critical of errors in communication when they are made by nonnative speakers.

Below are some suggestions to help mentors who have language or cultural differences from their mentees, particularly when the mentors are part of a minority culture.

- Be aware of the nuances of verbal and nonverbal communication in the setting in which mentoring or instruction will take place.
- Seek help from leaders at the school, native-speaking colleagues, or other professionals who have worked with similar groups of preservice teachers or mentees.
- Examine the meaning of mentoring in your culture before trying to understand mentees' needs, and encourage your mentees to do the same. Both parties should have a clear understanding about what to expect from the relationship.
- Provide mentees with a clear definition of their roles.
- Practice active listening, and provide continual feedback. These techniques help both parties in the relationship understand each other's cultures and perceptions of mentoring. In particular, mentors may ask mentees to complete reflective journal assignments and mentoring process evaluations.
- Work to establish trust with mentees so that they feel free to reflect honestly on the learning process.
- Let mentees hear you speaking in their native language. This strategy may eliminate the mystique associated with a second language and trigger mentees to appreciate your efforts to use their language.
- Do not become frustrated by cultural differences and cross-cultural mentoring challenges. Try to address and overcome your mentees' stereotypes by building trustworthy and caring relationships.

—*Fatma Aslan-Tutak and Adem Ekmekci*

Learning from a Novice Mentor's Mistakes

Years ago, during my second year as a high school mathematics teacher, I was asked to allow another teacher's student-teacher, Mary[1], to teach one of my classes. I gave Mary the option of choosing which class to teach, and she requested my last-period first-year-algebra class. I warned her that this class was a tough one and advised her to choose another period of the day, but

---

1. All names in the article are pseudonyms.

she insisted. I relented, as though my hands were tied. Mary started teaching my class a few days later and had difficulty with the students. I did not really know how to help her and, to my regret, basically left her to her own demise. In the end, Mary became something of a sacrificial lamb—she had a terrible experience and so did my students. I, on the other hand, returned to a group of students who were happier to have me as their teacher than they were before Mary showed up.

I learned two important lessons from this experience. First, I was completely unprepared to be Mary's mentor, and this deficiency had a significant negative impact on the experience for Mary, my students, and me. My lack of knowledge and training caused me to miss what could have been a valuable learning experience for all of us. Those who organize field experiences must ensure that mentors are equipped in all ways for their important role.

I also realized later that I had "used" Mary in some ways for my own selfish purposes. I wrote her off as a lost cause and waited in the wings, knowing that her experience would be negative and that my slight advantage in terms of teaching skill would cause me to look better when she left. I had the chance to reclaim my algebra students after their encounter with student-teaching, but I was unable to help Mary recover from her experience with my students. I wish I had been better prepared to support Mary, but more important, I wish I had given her the time and respect she deserved as a student, assistant, resource, and future colleague. I hope Mary has since forgiven me and that other novice mentors might learn from my mistakes.

—*Keith R. Leatham*

Mentoring Bilingual Mathematics Teachers

As a preservice mathematics teacher and English language learner, I knew mathematics better than English. I was terrified by the idea of stepping in front of a classroom where communicating mathematics was going to be the main part of my job. Becoming a teacher was my "vocación," and doing mathematics was my passion, but many times I wondered whether I was ever going to learn English well enough to teach my students effectively. The challenge of learning a new language took more than my own motivation and effort. It took human power to make me believe that it was possible. The source of that power was my mentors.

During my teacher-preparation program, my mentor recognized in me a natural mathematics talent, and saw my potential beyond the language impediment. However, he knew that I needed to become proficient in English, and he also knew how to help me. He had me write a master's thesis, make professional presentations, and participate in outreach activities at a local middle school. Although this process was long and painful, it connected teaching *and* mathematics—the two things I loved. At times I wondered if he really wanted to help or just make my life even more difficult. For example, I clearly remember him telling me, "I will not look at any written work if it has spelling errors" and "Rewrite the first five pages again, you are not thinking in English."

At the end of my schooling, I had written a master's thesis and knew how to implement a lesson for middle school students. Then came my first job . . . teaching precalculus to college freshmen. I was terrified. Yet one more time, another mentor stepped up to help guide me through those very crucial times. She asked me to prepare a lesson and deliver it to her and my colleagues a few days before class started. She and others gave me feedback and told me, "You don't have to talk all the time—write on the board the important stuff, make *them* talk; draw pictures; and write activities ahead of time." This was not all; I was required by my mentor to observe her class and others, as a way to learn how to teach this population of students. I observed more classes than my peers and was required to meet with her and talk about how things were going. Was I working harder than others? Of course I was. Was my situation fair? At the time I complained, and many times cried. Now I understand that my mentors cared deeply about my success as a professional. I had an impediment, and they were trying to help me overcome it by being forceful but reassuring.

Now, as a professor myself, I teach preservice teachers and train in-service teachers. I have a natural connection with bilingual teachers, and for some I have become their mentor. I have found no better way to give back what my mentors did for me.

— M. Alejandra Sorto

> **Building an Open Relationship: A Mentoring Vignette from Both Perspectives**

Michael: When Nicole approached me with serious concerns about her future as a teacher, I knew she might be facing a defining moment in her professional development. Spending the extra time to listen to her concerns initiated a mentoring relationship that has extended for the last three years.

Nicole: I first met Michael as a sophomore in his practicum in mathematics. Outside our classroom, Michael listened to my concerns about becoming a teacher, reassured me by showing respect for my concerns and sharing his experiences, and invited me to come back with further questions or problems. Michael's advice and support shaped who I am today, helping me to discover for myself the confidence to become a teacher. Michael also continued to support me through numerous other endeavors, including earning a summer job as a geometry teaching assistant for gifted students aged twelve to fourteen—an experience so rewarding that afterward, I had little doubt that I was meant to be a teacher—and acquiring a Fulbright scholarship to teach overseas.

As partners in a relatively successful mentoring relationship, we have identified specific characteristics and goals that helped us sustain an open and honest connection. These characteristics and goals are described in the following sections.

Michael's Input

- Respond quickly to mentees' needs. Time is a precious commodity, but change may take place slowly and individually. When confronted with scheduling choices, I always make time to mentor Nicole. I try to respond promptly to e-mail messages or phone calls and, if necessary, delay work on larger projects to speak with Nicole.

- Sharpen your listening skills. Although I have not always been a good listener, I have realized the need for improvement in this area in my role as a mentor. One way to become a better listener is to summarize the mentee's ideas before moving on to your own thoughts. Focusing on mentees' ideas helps these preservice teachers make the connection between theory and their own practice. Mentors should also avoid solving problems for their mentees, instead encouraging their protégés to come up with their own ideas.

- Show that you care. Mentors should truly care about the success of their mentees and make sure that the mentees are aware of that commitment. I sometimes find myself too quick to criticize and must actively choose to give positive feedback.

- Give specific feedback. Avoid general comments; more specific feedback helps the mentee identify meaningful core concepts. For example, I once told Nicole, "Your lesson plan has some nice closure questions. They help you check for understanding about the main ideas of prime numbers, factors, and composites and simultaneously ask your students how to win this mathematics game."

- Encourage growth. Mentors should push mentees to expand their horizons before they leave their undergraduate institutions. Conferences held by the National Council of Teachers of Mathematics, undergraduate research or curriculum opportunities, and grant or scholarship awards are among several options that help mentees establish their own professional development expectations.

Nicole's Input

- Pick someone you trust as a mentor. I knew Michael's teaching philosophy and identified him as a safe and reliable adviser. You will have different opinions than your mentor, but you must have the same general philosophy and goals in mind to have a successful mentorship. This trust is a foundation for a positive, long-term mentorship.

- Do not expect easy answers. Our mentoring relationship was one of the first I experienced in college. Initially, I expected Michael to interpret my interests and concerns and tell me which road to take. When he did not tell me exactly what to do, I was worried about the choices he left me to make. I later learned that the best advice guides you through your challenges instead of telling you exactly what to do.

- Trust yourself. Believing in myself was not always easy during the time when I was unsure of my goals for the future. Time and experience, however, helped me determine which directions to take in my career. I used some aspects of my mentor's advice, along with what I knew about myself, to guide my decisions. Putting the two perspectives together seemed to facilitate my decision making.

- Follow up with your mentor. Always let your mentor know how you have applied his or her advice and whether the advice has proved successful. Giving feedback on your mentor's advice allows the two of you to revisit topics of discussion and delve into them more deeply. Keep in mind that both immediate and long-term follow-up are important. Nearly three years after sitting in his class, I still write to Michael and tell him about my experiences as a new teacher. I know he appreciates hearing how his advice has influenced me long after I have left his classroom. Our ongoing dialogue has also allowed the relationship to develop from a mentorship to a professional association between colleagues.

    —*Michael E. Matthews and Nicole I. Guarino*

Two additional titles appear in the
Empowering the Mentor of the Mathematics Teacher series
(Gwen Zimmermann, series editor):

- ***Empowering the Mentor of the Beginning Mathematics Teacher,***
  edited by Gwen Zimmermann, Patricia Guinee, Linda M. Fulmore,
  and Elizabeth Murray

- ***Empowering the Mentor of the Experienced Mathematics Teacher,***
  edited by Gwen Zimmermann, Patricia Guinee, Linda M. Fulmore,
  and Elizabeth Murray

Please consult www.nctm.org/catalog for the availability of these titles,
as well as for a plethora of resources for teachers of mathematics
at all grade levels.

For the most up-to-date listing of NCTM
resources on topics of interest to mathematics
educators, as well as information on membership
benefits, conferences, and workshops, visit the
NCTM Web site at www.nctm.org.